Javier Caro Reina and Johannes Helmbrecht (Eds.)
Proper Names versus Common Nouns

Studia Typologica

Beihefte / Supplements
STUF – Sprachtypologie und Universalienforschung
　　　Language Typology and Universals

Editors
Thomas Stolz, François Jacquesson, Pieter C. Muysken

Editorial Board
Michael Cysouw (Marburg), Ray Fabri (Malta), Steven Roger Fischer (Auckland), Bernhard Hurch (Graz), Bernd Kortmann (Freiburg), Nicole Nau (Poznán), Ignazio Putzu (Cagliari), Stavros Skopeteas (Göttingen), Johan van der Auwera (Antwerpen), Elisabeth Verhoeven (Berlin), Ljuba Veselinova (Stockholm)

Volume 29

Proper Names versus Common Nouns

Morphosyntactic Contrasts in the Languages of the World

Edited by
Javier Caro Reina and Johannes Helmbrecht

DE GRUYTER
MOUTON

ISBN 978-3-11-163166-0
e-ISBN (PDF) 978-3-11-067262-6
e-ISBN (EPUB) 978-3-11-067274-9
ISSN 1617-2957

Library of Congress Control Number: 2022937124

Bibliographic information published by the Deutsche Nationalbibliothek
The Deutsche Nationalbibliothek lists this publication in the Deutsche Nationalbibliografie;
detailed bibliographic data are available on the Internet at http://dnb.dnb.de.

© 2024 Walter de Gruyter GmbH, Berlin/Boston
This volume is text- and page-identical with the hardback published in 2022.
Cover image: Alpha-C/iStock/Thinkstock

www.degruyter.com

Contents

Javier Caro Reina and Johannes Helmbrecht
Morphosyntactic contrasts between proper names and common nouns: an introduction —— 1

Corinna Handschuh
Personal names versus common nouns. Crosslinguistic findings from morphology and syntax —— 21

Javier Caro Reina
The definite article with personal names in Romance languages —— 51

Yves D'hulst, Rolf Thieroff and Trudel Meisenburg
River names. Definite articles and place names in West-Germanic and Romance —— 93

Johannes Helmbrecht
Proper names with and without definite articles: preliminary results —— 121

Elisheva Jeffay and Susan Rothstein
On personal names in construct states in Modern and Biblical Hebrew —— 155

Damaris Nübling
Von Heidel- nach Bamberg, von Eng- nach Irland? 'From Heidel- to Bamberg, from Eng- to Ireland?' On the delimitation of appellative proper names and genuine proper names —— 175

Iker Salaberri
D-marking on Basque personal names from a synchronic and diachronic perspective —— 205

Thomas Stolz and Nataliya Levkovych
On *Special Onymic Grammar (SOG)*: Definiteness markers in Fijian and selected Austronesian languages —— 237

Acknowledgements

The present volume goes back to the workshop *Proper names versus common nouns: morphosyntactic contrasts in the languages of the world* held on the 41st Annual Conference of the German Linguistic Society at the University of Bremen on March 06–08, 2019. We would like to express our gratitude to all speakers and participants of the workshop and to the authors for their contributions to this volume. We are indebted to the editors-in-chief – Thomas Stolz, Francois Jacquesson, and Pieter C. Muysken – for the admission of this volume into the *Studia Typologica* series and to the anonymous reviewers for their very helpful comments.

Javier Caro Reina and Johannes Helmbrecht

Javier Caro Reina and Johannes Helmbrecht
Morphosyntactic contrasts between proper names and common nouns: an introduction

Abstract: Proper names are considered as a special sub-category of nouns because they are used to refer to entities in the world and often behave grammatically like common nouns. This traditional view still prevails in general linguistics, linguistic typology, and descriptive approaches to individual languages. However, a closer look reveals that proper names can differ from common nouns with respect to morphological coding and syntactic/distributional behaviour. These morphological and syntactic contrasts are even more apparent among different proper name classes. We present here an introduction to the contrasts between proper names and common nouns, review current research on the morphosyntax of proper names, and summarize the main findings from the contributions of this volume.

Keywords: common nouns, morphosyntax, personal names, place names, proper names.

1 Introduction

Proper names (*nomina propria*) are linguistic expressions that are functionally defined by the following properties: Proper names have a direct, unique, and singular reference to an entity in the world (or rather: entity in the universe of discourse) such as individual persons, individual places, individual animals, individual cultural artifacts, individual historical or ritual events, etc. The reference of proper names is direct because it is not dependent on the semantic content of the proper name (as is the case with common nouns). Moreover, the reference of proper names is inherently specific and definite. That is, the speaker uses a proper name if the addressee is able to identify the referent of the proper name on the basis of their knowledge of the naming act. This naming act

---

Javier Caro Reina: Romanisches Seminar, Universität zu Köln, Albertus-Magnus-Platz, D-50923 Cologne. E-mail: javier.caroreina@uni-koeln.de.
Johannes Helmbrecht: Allgemeine und Vergleichende Sprachwissenschaft, Universität Regensburg, Universitätsstrasse 31, D-93053 Regensburg. E-mail: johannes.helmbrecht@ur.de.

may be a specific speech act of the type "X is called Y". Alternatively, the naming act remains implicit using words or complex expressions as proper names. This functional definition, which can be taken as a prototype definition in the sense of Croft (1991, 2001), significantly differs from the definition of common nouns. Common nouns (*nomina appellativa*) have semantic content and refer to classes of entities in the universe of discourse. In addition, they are not inherently are not inherently specific or definite.

Considering this fundamental functional difference between proper names and common nouns, it is reasonable to assume that there are differences in the morphosyntactic coding and distributional behaviour which can be explained in terms of the different semantic and pragmatic properties. Indeed, recent research on the morphosyntactic properties of proper names has shown that there are differences on all systemic levels of language — that is, on the phonetic/phonological, morphological, syntactic, and pragmatic level. Such formal differences are called dissociations (see Nübling 2005). The contributions in this volume deal in one way or another with such formal differences in certain domains of grammar.

Proper names have been traditionally the research object of onomastics, the linguistic sub-discipline that specifically deals with proper names. The main research interest of onomastics in the past and still up to the present has revolved around etymological and historical aspects of proper names. Studies on (synchronic) grammatical aspects of proper names (with or without comparison to common nouns) within this field are rare. As for linguistic typology, proper names have played an important role in scales (or hierarchies) such as the referentiality scale (Silverstein 1976: 113; Dixon 1979: 85; Croft 2003: 130).[1] The referentiality scale captures cross-linguistically recurrent patterns involving morphosyntactic phenomena such as split ergative case marking, differential object marking, etc. This scale predicts that categories ranking higher are more likely to be overtly marked than those ranking lower (see Bickel et al. 2014 for a critical view). Proper names occupy a prominent position between pronouns and common nouns. Notwithstanding, proper names have been regarded as a homogeneous group and have received little attention. For example, the *World Atlas of Language Structures* (WALS) studies the cross-linguistic patterns of morphosyntactic phenomena such as the definite article (Dryer 2013), locus of

---

[1] Other terms employed in the literature include activity scale, empathy hierarchy, extended animacy hierarchy, indexability hierarchy, and nominal hierarchy (see Haude and Witzlack-Makarevich 2016: 433 for references).

marking in possessive noun phrases (Nichols and Bickel 2013), etc. The focus, however, is on common nouns.

It is only recently that research on various aspects of the grammar of proper names has shown that proper names can differ from common nouns with respect to morphology as well as their syntactic behaviour and distribution (see, for instance, Wimmer 1973, Kalverkämper 1978, Kuhn and Serzisko 1982, Kolde 1995, Gallmann 1997, Anderson 2004, 2007, Nübling 2005, 2012, Van Langendonck 2007, Debus et al. 2014, Ackermann and Schlücker 2017, Helmbrecht et al. 2017, Dammel and Handschuh 2019, Helmbrecht 2020, Kempf et al. 2020, and Levkovych and Nintemann 2020). Nearly all these publications concentrate on European languages such as Germanic (especially English and German) and Romance languages (especially French, Italian, and Spanish). Only a very few articles (some of them in the edited volumes mentioned above) aim at a systematic typological comparison with respect to certain aspects of the grammatical properties of proper names (see Stolz et al. 2014, 2017 for case marking of place names, Handschuh 2017 for case and definiteness marking of proper names, Helmbrecht et al. 2018 for case marking of personal names in split ergative languages, Handschuh 2019 for sexus marking in personal names, and Mauri and Sansò 2019 for the associative plural marking on personal names). The lack of research on a comparative-typological level as well as on a language-specific level (in particular for Non-Indo-European languages) is surprising if we assume that proper names are one of the very few syntactic categories that are universal in the sense that they can be found in all languages (see, for instance, Hockett 1966).

In summary, previous work on the different behaviour of proper names and common nouns in different grammatical domains and in different languages clearly shows that the grammar of proper names constitutes a promising research area for linguistic typology, historical linguistics, and language variation. The goal of the present volume is to make a contribution to fill this gap of linguistic knowledge on the descriptive level and to strive for generalizations and their explanations. The volume brings together language-specific studies as well as cross-linguistic comparative studies on morphosyntactic differences between proper names and common nouns. All areas of grammatical marking are concerned, in particular nominal inflection such as case marking, definiteness marking, and marking of attributive relations.

2 Morphosyntactic phenomena

In this section, we will summarize the main findings of the contributions according to selected morphosyntactic phenomena: modification (2.1), case marking (2.2), the definite article (2.3), and possessive constructions (2.4).

2.1 Modification

Dixon (2010: 102, 108) observes that there are languages which avoid modification of proper names. For example, in Neverver, an Austronesian language spoken in Vanuatu, proper names cannot be modified by adjectives, demonstratives, numerals, possessives, relative clauses, and quantifiers (Barbour 2012: 87, 126).[2] Another example is modern Hebrew, where personal names cannot be pluralized or modified by relative clauses. By contrast, in Biblical Hebrew, personal names can be modified with a restrictive relative clause (see Jeffay and Rothstein this volume). Some languages allow modification of proper names, albeit with some restrictions. This is the case in Central Catalan, Icelandic, and Romanian, where personal names can be modified by an adjective only in postnominal position (see Caro Reina this volume for Central Catalan and Romanian).[3]

Other languages allow modification of proper names without restrictions. In this case, they can morphosyntactically behave like proper names or common nouns. An example of modified proper names which morphosyntactically pattern with proper names comes from Central Catalan and Balearic Catalan, where the proprial article *en* is retained with modification involving an adjective, a prepositional phrase, or a relative clause, as in the example:

(1) Central Catalan and Balearic Catalan
 en Joan / en Joan petit / en Joan amb barba / en Joan que ha respòs.
 'John / little John / the John with the beard / the John that answered.'

---

[2] Note that in Neverver, adjective modifiers are formed with uninflected stative verbs (see Barbour 2012: 300–301). In this respect, Barbour (2012: 118) talks about lexical modifiers.
[3] Julien (2005: 16–17) shows that in Icelandic, personal names differ from common nouns with respect to their position within the DP when they are modified by an adjective. Personal names occupy the initial position (*Anna litla* 'little Anna') as opposed to common nouns (*litla systir* 'little sister').

An example of modified proper names which morphosyntactically pattern with common nouns is Sinyar, a Central Sudanic language spoken in Chad and Sudan. In Sinyar, proper names and common nouns have different case endings (see Table 4). In the nominative singular, the personal name ˈBàkíítˈ 'Bakiit' takes the ending -lè while the common noun ʃékˈ 'chief' takes the ending -nì, as shown in (2). Crucially, Boyeldieu (2019: 484–485) points out that personal names are inflected like common nouns when they are modified by adjectives, demonstratives, numerals, and quantifiers (ˈBàkíît kwíí-ní 'another Bakiit').

(2) Sinyar (Boyeldieu 2019: 477, 479, 484)
 ˈBàkíítˈ-lè / ʃékˈ-nì / ˈBàkíît kwíí-ní
 Bakiit-NOM.SG / chief-NOM.SG / Bakiit another-NOM.SG
 'Bakiit / chief / another Bakiit'

Interestingly, in Icelandic, first names morphosyntactically pattern with personal names or common nouns depending on whether they are modified by an adjective or by a relative clause, respectively (Sigurðsson 2006: 226). Note that in Icelandic, first names optionally take the inflected preproprial article *hann* for the masculine (*hann Jón* 'John') or *hún* for the feminine (*hún María* 'Mary'). With regard to adjective modification, the preproprial article is retained. In contrast, with regard to relative clause modification, first names display the suffixed definite article *-an* typical of common nouns, as in the examples:

(3) Icelandic (Sigurðsson 2006: 226)
 hún María / hún María littla / þú ert fyrsta Marían sem ég kynnist
 'Mary / little Mary / you are the first Mary that I know'

2.2 Case marking

A closer look at languages which code the syntactic/semantic relations on the clause level reveals that proper names can morphosyntactically differ from common nouns in different ways. There are languages where proper names have specific case allomorphs, distinct case categories, or even a larger number of case categories. This is the case in Meryam Mir (Piper 1989: 31), Sinyar (Boyeldieu 2019), and Western Desert (Dixon 1980: 302) (see Handschuh this volume). In addition, the case marking pattern/alignment type changes if proper names occur as A or O arguments. This has been observed in Australian languages with split ergative case marking where personal names pattern sometimes with personal pronouns (nominative accusative marking pattern), sometimes with common nouns (ergative-absolutive marking pattern), and sometimes they show a kind of mixed behaviour with a so-called tripartite marking pattern. The former phenomenon can be observed in languages with differential object marking where personal names pattern with other human and definite NPs in O argument position overtly coded by means of an alternative case marker or a preposition.

Alternatively, proper names can display a smaller inflectional paradigm than common nouns. This is the case in German, which has experienced deflection (see Nübling 2012 and Ackermann 2018). Deflection contributes to the onymic schema constancy, according to which the shape of proper names is retained in order to enable their recognition and processing (Nübling 2005: 50–51). A further instance of deflection is found in Romanian (see Caro Reina this volume).

2.3 The definite article

If proper names are inherently definite and specific, it is reasonable to assume that proper names are not compatible with definiteness marking — that is, with definite articles. However, this is not necessarily true. In English, for instance, the definite article is disallowed with personal names. However, it is required with other proper name classes such as place names referring to mountains, lakes, rivers, etc. The same holds for standard German although in regional (or non-standard) varieties the definite article is almost the rule with first names and family names. On the other hand, there are languages such as modern Greek where personal names and place names obligatorily take the definite article. Up to now, there is no satisfying descriptive typology of the distribution of definite articles with proper names, nor are there satisfying explanations for the language-internal distribution of the definite article with different proper name classes. The contributions by Caro Reina, Handschuh, Helmbrecht, D'hulst et al., Nübling, Salaberri, and Stolz and Levkovych address the possible, obligatory, optional combination of the definite article with different proper name classes. In addition, special article forms which are exclusively used for proper names are investigated in the contributions by Caro Reina (Romance languages) and Stolz and Levkovych (Austronesian languages). All contributions demonstrate that there is a great deal of variation not only in single languages (and language varieties), but also in language families and cross-linguistically.

The use of the definite article with proper names has been related to the grammaticalization of the definite article (Greenberg 1978; Lyons 1999: 337; see

Szczepaniak and Flick 2020 for discussion). More recently, Löbner (2011) proposed a scale of uniqueness which captures the spread of the definite article from pragmatic to semantic definite contexts (see Ortmann 2014: 314 for a revised version). This scale predicts that the occurrence of the definite article with proper names implies its occurrence with individual and functional concepts. In addition, the presence of the definite article with personal pronouns implies the presence of the definite article with proper names.

(4) Scale of uniqueness (Ortmann 2014: 314, based on Löbner 2011)
deictic sortal noun < anaphoric sortal noun < sortal noun with establishing relative clause < definite associative anaphors < individual nouns/functional nouns < proper names < 3rd person personal pronouns < 2nd and 1st person personal pronouns

On this scale, proper names are viewed as a homogeneous group. A word of caution, however, is that place names can differ from personal names and, by extension, from personal pronouns. For example, the prediction that the occurrence of the definite article with personal pronouns implies the occurrence of the definite article with personal names and place names is borne out for Fijian and Nadrogā (see Stolz and Levkovych this volume). In contrast, this prediction is not borne out for Rapa Nui, where personal names and personal pronouns take the definite article *a* (*a Hotu* 'Hotu', *a ia* 's/he') while place names lack it (Ø *Rapa Nui* 'Rapa Nui') (Kieviet 2017: 102–103).

The *World Atlas of Language Structures* (WALS) contains a chapter dedicated to the definite article with common nouns (Dryer 2013). However, little is known about the occurrence of the definite article with proper names (or proper name classes) from a cross-linguistic perspective. We will now outline the four possible combinations of the presence (or absence) of the definite article with proper names and common nouns which can be found cross-linguistically. For example, there are languages without definite articles. One example is Manam, an Austronesian language spoken in the northeast of New Guinea, which lacks the definite article, both with proper names and with common nouns (Lichtenberk 1983: 523).

(5) Manam (Lichtenberk 1983: 523)
Ø manabumbía / Ø iribóaba / Ø tamóata / Ø áine
'Manabumbia / Iriboaba / (the) man / (the) woman'

In some languages, the definite article is restricted to common nouns, as in Maṽea, an Austronesian Language spoken in Vanuatu, where personal names lack the definite article while common nouns are optionally followed by the definite article *le* (Guérin 2011: 48, 149), as shown in (6).[4] In other languages, the definite article is restricted to proper names. This is the case in Butam (Laufer 1959: 199–200), Taulil (or Tulil) (Laufer 1950: 634; Meng 2018: 11, 116, 119), and Toqabaqita (Lichtenberk 2008: 47, 50, 237). For example, in Toqabaqita, an Austronesian language spoken in the Solomon Islands, masculine and feminine personal names are optionally accompanied by *tha* and *ni* respectively while common nouns lack the definite article (Lichtenberk 2008: 237), as illustrated in (7). The patterns found in these languages challenge the grammaticalization pathway of the definite article according to which the definite article is found with common nouns prior to proper names.

(6) Maṽea (Guérin 2011: 48, 193, 375)
Ø *Moltas* / Ø *Vovrodal* / *tamlo le* / *ṽaṽina le*
'Moltas / Vovrodal / the man / the woman'

(7) Toqabaqita (Lichtenberk 2008: 47, 50, 237)
tha Ulufaalu / *ni Saelifiqa* / Ø *wane* / Ø *kini*
'Ulufaalu / Saelifiqa / the man / the woman'

Finally, there are languages which exhibit the definite article both with proper names and common nouns, as in Kove and Vitu, two Austronesian languages spoken in West New Britain. For example, Vitu employs different definite articles for proper names and common nouns (see Berg and Bachet 2006: 27 for details). More specifically, *a* occurs with personal names and place names while *na* occurs with common nouns.

(8) Vitu (Berg and Bachet 2006: 27–30, 33–35)
a Kalago / *a Lama* / *na tamohane* / *na malala*
'Kalago / Lama / the man / the village'

The patterns of the definite article with proper names and common nouns in these languages are summarized in Table 1. However, the question remains open as to how these patterns are distributed cross-linguistically. In this re-

---

4 As pointed out by Guérin (2007: 549–550), the definite article *le* is in a transitional grammaticalization stage, where it is developing from pragmatic to semantic definite contexts.

spect, first typological surveys have been carried out by Handschuh (2017). In addition, this volume contains three papers dealing with the occurrence of the definite article with personal names and place names from a typological perspective (Caro Reina; Helmbrecht; Stolz and Levkovych).

Table 1: Definite article with proper names and common nouns in selected languages.

Language	Proper name	Common noun	Reference
Manam	–	–	Lichtenberk (1983)
Mav̌ea	–	+	Guérin (2011)
Butam, Taulil, Toqabaqita	+	–	Laufer (1950; 1959), Lichtenberk (2008), Meng (2018)
Kove, Vitu	+	+	Berg and Bachet (2006), Sato (2013)

Table 2 summarizes the findings of the papers by Caro Reina, Helmbrecht, and Salaberri, which show that the occurrence of the definite article with personal names can be motivated by semantic-pragmatic, lexical, morphosyntactic, phonological, and sociolinguistic factors. Importantly, the use of the definite article can result from one single factor or from the combination of several factors. For example, in some varieties of Basque, the definite article occurs with feminine hypocoristics. In European Portuguese, the definite article is employed in colloquial register with ordinary names and with negatively connoted famous names.

Table 2: Factors conditioning the occurrence of the definite article with personal names (see Caro Reina this volume, Helmbrecht this volume, and Salaberri this volume).

Factor	Language
Semantic/pragmatic	
Ordinary name	Balearic Catalan, Central Catalan, Galician, Italian, Brazilian Portuguese, European Portuguese, Occitan, Sursilvan, varieties of French/German/Italian/Spanish
Famous name	Asturian, Balearic Catalan, French, Galician, Italian, Brazilian Portuguese, Sardinian, Spanish, Sursilvan
Negative connotation	Asturian, Galician, European Portuguese, varieties of German/Spanish
Lexical	
Personal name type	Varieties of Catalan
Personal name class	Arabic, Asturian, Breton, Bulgarian, Dime, Galician, Italian, Kabardian, Sursilvan, varieties of Basque/Spanish
Foreignness (native vs. borrowed names)	Hidatsa

Morphosyntactic	
Case	Galician, Occitan, Romanian
Gender	Asturian, Italian, Occitan, Savosavo, Spanish, Sursilvan, varieties of Basque/Catalan/Italian/Spanish
Phonological	
Word-initial segment	Roussillon Catalan
Word-final segment	Romanian
Sociolinguistic	
Non-standard	French, German, Hungarian, Italian, Spanish
Style	Balearic Catalan, Central Catalan, Brazilian Portuguese, European Portuguese, Romanian

Helmbrecht (this volume) addresses the question of whether there is an implication regarding the presence of the definite article with personal names and place names. He concludes that there is no such implication since all four combinations are possible: definite article with personal names and place names (Qaqet), definite article only with personal names (Tepehuan), definite article only with place names (Hungarian), and definite article absent from personal names and place names (Tuvaluan). This suggests that the emergence of the definite article with personal names and place names obeys different strategies: a pragmatic one with personal names, but a semantic one with place names (that is, from less prototypical to more prototypical) (see Caro Reina 2020 for the definite article with place names). Crucially, the occurrence of the definite article is sensitive to the class of personal name (first name, family name, etc.) and place name (city name, country name, etc.). In addition, it can be constrained in prepositional phrases (see Himmelmann 1998: 323–338 for common nouns in Albanian, Nkore-Kiga, Romanian, Tagalog, and Germanic languages). In Fijian, for example, article drop applies to proper names in spatial relations (see Stolz and Levkovych this volume). In Galician and Occitan, the *a*-marker favours the absence of the definite article with personal names (see Caro Reina this volume). In French, Italian, Portuguese, Romanian, Sardinian, and Sursilvan, we find article drop with place names in prepositional phrases (see Caro Reina 2020: 36–40).

The presence of the definite article with proper names and common nouns raises the question of whether we can talk about a morphosyntactic contrast at all. In languages where the definite article is completely grammaticalized (that is, it occurs with all personal names and place names), we could argue that there is no difference. However, in languages with sex-based gender systems, the definite article functions rather as a classifier for specific proper name subclasses (see Nübling 2020 for German). By contrast, in languages where the definite article is not grammaticalized, the presence of the definite article with personal names is pragmatically motivated (connotation, familiarity, etc.). In this respect, the definite article fulfills different functions with personal names and common nouns.

2.4 Possessive constructions

Proper names can differ from common nouns with respect to possessive noun phrases where proper names appear as head nouns (possessors). The differences involve word order (see Nübling 2017: 355–356 for pre- and postnominal genitives in German), possessive markers (see Stolz et al. 2017: 125–130 for Faroese, Icelandic, and Romanian), and locus of marking. With regard to possessive markers, in Romanian, personal names are preceded by the proprial article *lui* while place names and common nouns are inflected by means of the suffixed definite article (see Caro Reina this volume). In addition, ordinary and famous names behave differently since ordinary names can take the proprial article *lui* or the preposition *de* 'of' (*Romanele lui/de Ion* 'the novels of Ion') while names of famous writers can only take the preposition *de* 'of' (*Romanele *lui/de Rebreanu* 'the novels of Rebreanu').

(9) Romanian
    lui Ion / Bucureşti-ului / băiat-ului
    'of John / of Bucharest / of the boy'

With regard to locus of marking, Nichols and Bickel (2013) distinguish between head marking, dependent marking, double marking, and zero marking (or juxtaposition). For example, French is a dependent marking language where the possessor noun carries the marker. This applies for personal names (*la maison de Jean* 'the house of John') and common nouns (*la maison du maire* 'the house of the mayor'). By contrast, Old French exhibits juxtaposition with personal names (and functional nouns), but dependent marking by means of the prepositions *à/de* with common (sortal) nouns (Buridant 2000: 99–100), as can be seen in (10). This stage is retained in conservative varieties of French such as Picard and Walloon. An example from Picard is given in (11).

(10) Old French (Palm 1977: 69, 121)
    li filz Ø Lancelot / le fil au chevalier
    'the son of Lancelot / the son of the knight'

(11) Picard (Haigneré 1901: 254–255)
    la maison Ø Jean / le ferme du maire

'the house of John / the farm of the mayor'

Another example is modern Hebrew (see Jeffay and Rothstein this volume). In Modern Hebrew there are three possessive constructions: the so-called construct state, the free genitive with the preposition *šel* (שֶׁל), and double marking (see Danon 2017: 64 for a corpus analysis). Construct states involve head marking structures where the head noun (possessee) is morphologically inflected for the construct state while the dependent (possessor) appears in the absolute form. The peculiarity of these constructions is that personal names cannot occur as possessor, as in (12). Interestingly, this restriction was absent from Biblical Hebrew.

(12) Modern Hebrew
    bet   *ha-mora*   /   **ariela*
    house.M.SG.CS   DEF-teacher.F.SG.ABS /   Ariella
    'the house of the teacher / Ariella'

3 Morphosyntactic differences between proper name classes and common nouns

The papers in this volume have shown not only that proper names can morphosyntactically differ from common nouns, but also that proper name classes can behave differently. In this section, we will classify these morphosyntactic contrasts into four types: (a) personal names and place names differ from common nouns (3.1); (b) personal names deviate from place names and common nouns (3.2); (c) place names behave differently from personal names and common nouns (3.3); and (d) personal names, place names, and common nouns exhibit distinct morphosyntactic patterns (3.4). This implies that place names can morphosyntactically pattern either with personal names or with common nouns. Alternatively, they can morphosyntactically deviate from personal names and common nouns. In this respect, Stolz and Levkovych (this volume) talk about specific onymic grammar (SOG), specific anthroponymic grammar (SAG), and specific toponymic grammar (STP).

3.1 Personal names and place names vs. common nouns

In some languages, proper names (both personal names and place names) morphosyntactically differ from common nouns. Evidence comes from definite articles in Fijian and Maori, article-drop in prepositional phrases in Fijian and Maori, inflection in Sinyar, and spatial relations in Fijian (see Handschuh this volume and Stolz and Levkovych this volume). In the Austronesian languages Fijian and Maori, we find different article forms for proper names and common nouns, as shown in Table 3.

Table 3: Marker systems in Fijian and Maori (Stolz and Levkovych this volume).

Language	Personal name	Place name	Common noun
Fijian	(k)o	(k)o	na
Maori	a	a	te

In addition, in Fijian, article-drop in prepositional phrases is sensitive to proper names and common nouns. Proper names, which take the definite article *(k)o*, experience article-drop in spatial relations involving locative, ablative, and directive. This does not hold for common nouns, which retain the definite article *na*. Similarly, in Maori a set of prepositions (other than *i, ki, kei, hei*) trigger article-drop in prepositional phrases with proper names, but not with common nouns.

In Sinyar, a Central Sudanic language spoken in Chad and Sudan, proper names considerably differ from common nouns with respect to inflection (see Table 4). First, proper names overtly code more cases (nominative, genitive, accusative, and adverbial) than common nouns (nominative and adverbial). Second, proper names and common nouns mostly display different case allomorphs.

Table 4: Inflection of proper names and common nouns in Sinyar (Boyeldieu 2019: 483).

Case	Number	Proper name	Common noun
Nominative	Singular	-n/(ʼ)-ǹ/-lè	-n/(ʼ)-ǹ/-Ní/(ʼ)-Nì
	Plural	-ngè	-sí/(ʼ)-sì
Genitive	Singular	-nàʲ/-nâʲ	
	Plural	-ngè	
Accusative	Singular	-(y)àà	
	Plural	-ngàá	
Adverbial	Singular	°-lèè	-tí/(ʼ)-tì
	Plural	°-ngèèr	

A further example are Bantu languages, where personal names and place names are assigned to class 1a while common nouns are assigned to different noun classes. Class 1a is characterized by the absence of classifiers (see Van de Velde 2019: 240–241).

3.2 Personal names vs. place names and common nouns

Evidence for personal names behaving differently from place names and common nouns can be found in possessive constructions in modern Hebrew and inflection in Romanian (see Caro Reina this volume and Jeffay and Rothstein this volume). In modern Hebrew, construct states can contain place names and common nouns, but not personal names.

(13) Hebrew
    a. *tošvey*　　　　　　*tel Aviv*
       resident.M.PL.CS　　Tel Aviv
       '(the) residents of Tel Aviv'
    b. *bet*　　　　　　*ha-mora*
       house.M.SG.CS　　DEF-teacher.F.SG.ABS
       'the house of the teacher'

In Romanian, personal names are inflected differently from place names and common nouns. More specifically, personal names are uninflected in the nominative-accusative while they take the proprial article *lui* in the genitive-dative. By contrast, place names and common nouns require the suffixed definite article for the nominative-accusative and genitive-dative, as illustrated in (14).

(14) Romanian
    a. *IonØ / Bucureşti-ul / băiat-ul*
       'John / Bucharest / the boy'
    b. *lui Ion / Bucureşti-ului / băiat-ului*
       'of/to John / of/to Bucharest / of/to the boy'

We can add Butam and Taulil (or Tulil), two Papuan languages spoken in East New Britain, where the definite article is restricted to personal names (see Section 2.3).[5]

(15) Taulil (Meng 2018: 64, 116, 120, 161)
to Rikie / e Luisa / Ø Nolvon / Ø lok / Ø vakue
'Rikie / Luisa / Nolvon / (the) man / (the) woman'

3.3 Place names vs. personal names and common nouns

Evidence for place names behaving differently from personal names and common nouns is derived from article-drop in prepositional phrases in Maori and definite article in Southeastern Tepehuan (see Stolz and Levkovych this volume and Handschuh this volume). In Maori, a set of prepositions (*i, ki, kei, hei*) trigger article-drop with place names, but not with personal names and common nouns, which retain the definite article. In Southeastern Tepehuan, a Uto-Aztecan language spoken in northern Mexico, the definite article *gu* occurs with personal names and common nouns, but not with place names.

(16) Southeastern Tepehuan (Willett 1991: 80, 206)
gu Juan / Ø Susba'n-tam / gu chio'ñ
'John / Frog Town / the man'

An additional example is Kove, an Austronesian language spoken in West New Britain, where the definite article *to* occurs with personal names, place names, and common nouns. However, *to* follows personal names and common nouns, but precedes place names (Sato 2013: 133–134).

(17) Kove (Sato 2013: 133–134)
Paul to / to Kimbe / pana to
'Paul / Kimbe / the people'

---

5 In Taulil, in addition to personal names, the definite article *a* occurs with loanwords such as *a-pos* 'post' and *a-purpur* 'flower', which are borrowed from Tok Pisin *pos* and Tolai *purpur*, respectively (Meng 2018: 116).

3.4 Personal names vs. place names vs. common nouns

Evidence for personal names, place names, and common nouns behaving differently from each other comes from gender in Bukiyip and definite articles in Chamorro, Nadrogā, Rapa Nui, and Taiof (see Handschuh this volume and Stolz and Levkovych this volume). Bukiyip, a Torricelli language spoken in Papua New Guinea, exhibits 18 noun classes. Personal names are assigned to class 18 while place names are assigned to class 17. With regard to common nouns, class 4 contains female persons (*élmatok* 'woman') while class 7 contains male persons (*élman* 'man').

In Rapa Nui, an Oceanic language spoken on Easter Island, personal names take the definite article *a* while common nouns take the definite article *te*. By contrast, place names lack the definite article (Kieviet 2017: 101–103). One exception are place names metonymically employed to refer to their inhabitants. In this case, place names take the definite article typical of personal names (*a Rapa Nui* 'the people of Rapa Nui') (Kieviet 2017: 103).

(18) Rapa Nui (Kieviet 2017: 101–102)
 a Hotu / Ø Rapa Nui / te taŋata
 'Hotu / Rapa Nui / the man'

In the Austronesian languages Chamorro, Nadrogā, and Taiof, we find different article forms for personal names, place names, and common nouns, as shown in Table 5.

Table 5: Marker systems in Chamorro, Nadrogā, and Taiof (Stolz and Levkovych this volume).

Language	Personal name	Place name	Common noun
Chamorro	si / as	iya	i / ni ~ nu
Nadrogā	o	i	na
Taiof	e	Ø	a / i

We can add Tolai, an Austronesian language spoken in East New Britain, where the definite articles *to* and *la* occur with traditional male and female names respectively while the definite article *a* occurs with common nouns (Mosel 1984: 16–17). In contrast, place names do not require the definite article unless they are employed metonymically to refer to the inhabitants of the place. In this case, they take the definite article typical of common nouns (*a Bailing* 'the Bailing people') (Mosel 1984: 80).

(19) Tolai (Mosel 1984: 16–17)
 to Vuvu / Ø Bailing / a tutana
 'Vuvu / Bailing / the man'

Abbreviations

ABS	absolute
CS	construct state
DEF	definite article
F	feminine
M	masculine
NOM	nominative
PL	plural
SG	singular

References

Ackermann, Tanja. 2018. *Grammatik der Namen im Wandel. Diachrone Morphosyntax der Personennamen im Deutschen* (Studia Linguistica Germanica 134). Berlin & Boston: de Gruyter.
Ackermann, Tanja & Barbara Schlücker (eds.). 2017. The morphosyntax of proper names. [Special issue]. *Folia Linguistica* 51(2).
Anderson, John M. 2004. On the grammatical status of names. *Language* 80(3). 435–474.
Anderson, John M. 2007. *The grammar of names*. Oxford: Oxford University Press.
Barbour, Julie. 2012. *A grammar of Neverver* (Mouton Grammar Library 60). Berlin & Boston: Walter de Gruyter.
Berg, René van den & Peter Bachet. 2006. *Vitu grammar sketch*. Ukarumpa, EHP: SIL Printing Press.
Bickel, Balthasar, Alena Witzlack-Makarevich & Taras Zakharko. 2014. Typological evidence against universal effects of referential scales on case alignment. In Ina Bornkessel-Schlesewsky, Andrej Malchukov & Marc Richards (eds.), *Scales and hierarchies. A cross-disciplinary perspective* (Trends in Linguistics. Studies and Monographs 277), 7–44. Berlin: de Gruyter.
Boyeldieu, Pascal. 2019. Proper names and case markers in Sinyar (Chad/Sudan). In Antje Dammel & Corinna Handschuh (eds.), *Grammar of names*. [Special issue]. *Language Typology and Universals* 72(4). 467–503.
Buridant, Claude. 2000. *Grammaire nouvelle de l'ancien français*. Paris: SEDES.
Caro Reina, Javier. 2020. The definite article with place names in Romance languages. In Nataliya Levkovych & Julia Nintemann (eds.), *Aspects of the grammar of names: Empirical case studies and theoretical topics* (LINCOM Studies in Language Typology 33), 25–51. München: LINCOM.

Croft, William. 1991. *Syntactic categories and grammatical relations: The cognitive organization of information*. Chicago: University of Chicago Press.
Croft, William. 2001. *Radical construction grammar: Syntactic theory in typological perspective*. Oxford: Oxford University Press.
Croft, William A. 2003. *Typology and universals*. Cambridge: Cambridge University Press.
Dammel, Antje & Corinna Handschuh (eds.). 2019. Grammar of names. [Special Issue]. *Language Typology and Universals* 72(4).
Danon, Gabi. 2017. Imagine no possession: John Lennon in the construct state. In Noa Brandel (ed.), *Proceedings of Israel Association for Theoretical Linguistics* 33, 49–68 (MIT Working Papers in Linguistics 89). https://www.iatl.org.il.
Debus, Friedhelm, Rita Heuser & Damaris Nübling (eds.). 2014. *Linguistik der Familiennamen* (Germanistische Linguistik 225–227). Hildesheim: Olms.
Dixon, Robert M. W. 1979. Ergativity. *Language* 55(1): 59–138.
Dixon, Robert M.W. 1980. *The languages of Australia*. Cambridge: Cambridge University Press.
Dixon, Robert M. W. 2010. *Basic linguistic theory. Volume 1: Methodology*. Oxford: Oxford University Press.
Dryer, Matthew S. 2013. Definite articles. In Matthew S. Dryer & Martin Haspelmath (eds.), *The world atlas of language structures online*. Leipzig: Max Planck Institute for Evolutionary Anthropology. http://wals.info/chapter/37 (checked 15.03.2022).
Gallmann, Peter. 1997. Zur Morphosyntax der Eigennamen im Deutschen. In Elisabeth Löbel & Gisa Rauh (eds.), *Lexikalische Kategorien und Merkmale*, 73–86. Tübingen: Niemeyer.
Greenberg, Joseph H. 1978. How does a language acquire gender markers? In Joseph H. Greenberg, Charles A. Ferguson & Edith A. Moravcsik (eds.), *Universals of human language. Volume 3: Word structure*, 47–82. Stanford, CA: Stanford University Press.
Guérin, Valérie. 2007. Definiteness and specificity in Mav̄ea. *Oceanic Linguistics* 46(2). 538–553.
Guérin, Valérie. 2011. *A grammar of Mav̄ea: An Oceanic language of Vanuatu*. Honolulu: University of Hawai'i Press.
Haigneré, Daniel. 1901. *Le patois boulonnais comparé avec les patois du nord de la France. Introduction, phonologie, grammaire*. Paris: A. Picard.
Handschuh, Corinna. 2017. Nominal category marking on personal names. A typological study of case and definiteness. *Folia Linguistica* 51(2). 483–504.
Handschuh, Corinna. 2019. The classification of personal names: A crosslinguistic study of sex-specific forms, classifiers and gender marking on personal names. In Antje Dammel & Corinna Handschuh (eds.), *Grammar of names*. [Special issue]. *Language Typology and Universals* 72(4). 539–572.
Haude, Katharina & Alena Witzlack-Makarevich. 2016. Referential hierarchies and alignment: An overview. *Linguistics* 54 (3), 433–441.
Helmbrecht, Johannes & Corinna Handschuh. 2017. Regeln zur Bildung von Personennamen. Eine typologische Studie. Regensburg: University of Regensburg manuscript.
Helmbrecht, Johannes, Lukas Denk, Sarah Thanner & Ilenia Tonetti. 2018. Morphosyntactic coding of proper names and its implications for the Animacy Hierarchy. In Sonja Cristofaro & Fernando Zúñiga (eds.), *Typological hierarchies in synchrony and diachrony* (Typological Studies in Language 121), 377–402. Amsterdam & Philadelpia: John Benjamins.
Helmbrecht, Johannes. 2020. Form and function of personal names: dimensions of the morphosyntactic diversity. In Nataliya Levkovych & Julia Nintemann (eds.), *Aspects of the grammar of names. Empirical case studies and theoretical topics* (LINCOM Studies in Language Typology 33), 1–25. München: LINCOM.

Helmbrecht, Johannes, Damaris Nübling & Barbara Schlücker (eds.). 2017. *Namengrammatik* (Linguistische Berichte – Sonderheft 23). Hamburg: Buske.
Himmelmann, Nikolaus P. 1998. Regularity in irregularity: Article use in adpositional phrases. *Linguistic Typology* 2(3). 315–353.
Hockett, Charles F. 1966. The problem of universals in language. In Joseph H. Greenberg (ed.), *Universals of language*, 1–29. 2nd edn. Cambridge, MA: The MIT Press.
Julien, Marit. 2005. *Nominal phrases from a Scandinavian perspective* (Linguistik Aktuell/Linguistics Today 87). Amsterdam & Philadelphia: John Benjamins.
Kalverkämper, Hartwig. 1978. *Textlinguistik der Eigennamen*. Stuttgart: Klett-Cotta.
Kempf, Luise, Damaris Nübling & Mirjam Schmuck (eds.). 2020. *Linguistik der Eigennamen* (Linguistik, Impulse & Tendenzen 88). Berlin: De Gruyter.
Kieviet, Paulus. 2017. *A grammar of Rapa Nui*. Berlin: Language Science Press.
Kolde, Gottfried. 1995. Grammatik der Eigennamen (Überblick). In Ernst Eichler et al. (eds.), *Namenforschung. Ein internationales Handbuch zur Onomastik* (Handbücher zur Sprach- und Kommunikationswissenschaft 11), vol. 1, 400–408. Berlin & New York: Walter de Gruyter.
Kuhn, Wilfried & Fritz Serzisko. 1982. Eigennamen im Rahmen der Dimension der Apprehension. In Hansjakob Seiler & Christian Lehmann (eds.), *Apprehension: Das sprachliche Erfassen von Gegenständen. Teil 1: Bereich und Ordnung der Phänomene*, 277–293. Tübingen: Narr.
Laufer, P. Carl. 1950. Die Taulil und ihre Sprache auf Neubritannien. *Anthropos* 45(4–6). 627–640.
Laufer, P. Carl. 1959. P. Futschers Aufzeichnungen über die Butam-Sprache (Neubritannien). *Anthropos* 54(1–2). 183–212.
Levkovych, Nataliya & Julia Nintemann (eds.). 2020. *Aspects of the grammar of names: Empirical case studies and theoretical topics* (LINCOM Studies in Language Typology 33). München: LINCOM.
Lichtenberk, Frantisek. 1983. *A grammar of Manam* (Oceanic Linguistics Special Publications 18). Honolulu: University of Hawai'i Press.
Lichtenberk, Frantisek. 2008. *A grammar of Toqabaqita* (Mouton Grammar Library 42). Berlin & Boston: Walter de Gruyter.
Löbner, Sebastian. 2011. Concept types and determination. *Journal of Semantics* 28(3). 279–333.
Lyons, Christopher. 1999. *Definiteness*. Cambridge: Cambridge University Press.
Mauri, Caterina & Andrea Sansò. 2019. *Nouns & co.* Converging evidence in analysis of associative plurals. In Antje Dammel & Corinna Handschuh (eds.), Grammar of names. [Special issue]. *Language Typology and Universals* 72(4). 603–626.
Meng, Chenxi. 2018. *A grammar of Tulil*. Melbourne: La Trobe University PhD thesis.
Mosel, Ulrike. 1984. *Tolai syntax and its historical development* (Pacific Linguistics B-92). Canberra: Australian National University.
Nichols Johanna & Balthasar Bickel. 2013. Locus of marking in possessive noun phrases. In Matthew S. Dryer & Martin Haspelmath (eds.), *The world atlas of language structures online*. Leipzig: Max Planck Institute for Evolutionary Anthropology. http://wals.info/chapter/37 (checked 15.03.2022).
Nübling, Damaris. 2005. Zwischen Syntagmatik und Paradigmatik: Grammatische Eigennamenmarker und ihre Typologie. *Zeitschrift für Germanistische Linguistik* 33(1). 25–56.
Nübling, Damaris. 2012. Auf dem Weg zu Nicht-Flektierbaren: Die Deflexion der deutschen Eigennamen diachron und synchron. In Björn Rothstein (ed.), *Nicht-flektierende Wortarten* (Linguistik - Impulse und Tendenzen 47), 224–246. Berlin & New York: Walter de Gruyter.

Nübling, Damaris. 2017. The growing distance between proper names and common nouns in German: On the way to onymic schema constancy. In Tanja Ackermann & Barbara Schlücker (eds.), *The morphosyntax of proper names*. [Special issue]. *Folia Linguistica* 51(2). 341–367.

Nübling, Damaris. 2020. *Die Capital – der Astra – das Adler*: The emergence of a classifier system for proper names in German. In Renata Szczepaniak & Johanna Flick (eds.), *Walking on the grammaticalization path of the definite article: Functional main and side roads* (Studies in Language Variation 23), 228–249. Amsterdam & Philadelphia: John Benjamins.

Ortmann, Albert. 2014. Definite article asymmetries and concept types: Semantic and pragmatic uniqueness. In Thomas Gamerschlag, Doris Gerland, Rainer Osswald & Wiebke Petersen (eds.), *Frames and concept types: Applications in language and philosophy* (Studies in Linguistics and Philosophy 94), 293–321. Cham: Springer.

Palm, Lars 1977. *La construction "li filz le rei" et les constructions concurrentes avec "a" et "de" étudiées dans les oeuvres littéraires de la seconde moitié du XIIe siècle et du premier quart du XIIIe siècle.* Uppsala: Almqvist & Wiksell.

Piper, Nick. 1989. *A sketch grammar of Meryam Mir*. Canberra: Australian National University MA thesis.

Sato, Hiroko. 2013. *Grammar of Kove: An Austronesian language of the West New Britain province, Papua New Guinea*. Hawai'i: University of Hawai'i PhD thesis.

Sigurðsson, Halldór Ármann. 2006. The Icelandic noun phrase: Central traits. *Arkiv för nordisk filologi* 121. 193–236.

Silverstein, Michael. 1976. Hierarchy of features and ergativity. In Robert M. W. Dixon (ed.), *Grammatical categories in Australian languages*, 112–171. Atlantic Highlands, NJ: Humanities Press.

Stolz, Thomas, Lestrade Sander & Christel Stolz. 2014. *The cross-linguistics of zero-marking of spatial relations* (Studia Typologica. Supplements STUF - Language Typology and Universals 15). Berlin & Boston: De Gruyter.

Stolz, Thomas, Nataliya Levkovych & Aina Urdze. 2017. Die Grammatik der Toponyme als typologisches Forschungsfeld. In Johannes Helmbrecht, Damaris Nübling & Barbara Schlücker (eds.), *Namengrammatik* (Linguistische Berichte – Sonderheft 23), 121–146. Hamburg: Buske.

Szczepaniak, Renata & Johanna Flick. 2020. Introduction. In Renata Szczepaniak & Johanna Flick (eds.), *Walking on the grammaticalization path of the definite article: Functional main and side roads* (Studies in Language Variation 23), 1–13. Amsterdam & Philadelphia: John Benjamins.

Van de Velde, Mark. 2019. Nominal morphology and syntax. In Mark Van de Velde, Koen Bostoen, Derek Nurse & Gérard Philippson (eds), *The Bantu languages*, 237–269. 2nd edn. London & New York: Routledge.

Van Langendonck, Willy. 2007. *Theory and typology of proper names*. Berlin & New York: Mouton de Gruyter.

Willett, Thomas L. 1991. *A reference grammar of Southeastern Tepehuan*. Dallas, TX: Summer Institute of Linguistics.

Wimmer, Rainer. 1973. *Der Eigenname im Deutschen. Ein Beitrag zu seiner linguistischen Beschreibung* (Linguistische Arbeiten 11). Tübingen: Niemeyer.

Corinna Handschuh
Personal names versus common nouns

Crosslinguistic findings from morphology and syntax

Abstract: When comparing the morphosyntactic properties of proper names and common nouns, the latter have often been argued to have a richer linguistic structure, e.g. with respect to their inflectional potential, while names have been viewed as defective in some respect. The paper investigates differences in the encoding of inflectional categories as well as in the syntactical behaviour between common nouns and proper names, while also taking differences between personal names and place names into consideration. It is demonstrated that names have unique grammatical properties and sometimes even have a larger inventory of forms than common nouns.

Keywords: common nouns, modifiers, morphosyntax, nominal categories, proper names.

1 What's in a name?

The difference between a name and a noun is blurred in everyday usage as well as in a large portion of linguistic work.[1] This does not come as a surprise since the Latin word *nomen* encompasses both meanings, and after all linguistic terminology in the western world has been hugely influenced by Latin. However, the morpho-syntactic properties of names and nouns vary to a greater or lesser extent depending on the language one investigates.[2] Although, these differences have occasionally been noticed and commented on, e.g. by Sasse (1993: 195), who concluded that they should be treated as distinct parts of speech (at

---

[1] For instance, Schachter and Shopen in their paper on parts-of-speech systems refer to the "traditional definition of nouns, assigning the label *noun* to the class of words in which occur the names of most persons, places, and things" (2007: 5).
[2] It is, of course, also possible for a language not to make any distinction between proper names and common nouns in their morpho-syntactic behaviour, as pointed out by an anonymous reviewer, who suggested Japanese as such a language.

Corinna Handschuh: Allgemeine und Vergleichende Sprachwissenschaft, Universität Regensburg, Universitätsstr. 31, 93053 Regensburg, Germany, corinna.handschuh@ur.de

https://doi.org/10.1515/9783110672626-002

least in some languages) due to those differences, more detailed studies of the morphosyntax of proper nouns have only been undertaken rather recently. Starting from Van Langendonck's (2007) and Anderson's (2007) monographs, a number of publications dedicated to the crosslinguistic comparison of common nouns and proper names have been published in the last few years (e.g. the volumes edited by Helmbrecht et al. 2017, Ackermann and Schlücker 2017, and Dammel and Handschuh 2019, to name just a few).

Linguistic work that has made a distinction between names and nouns has often focussed on the defectiveness of names (cf. Allerton 1987: 81), i.e. for instance their limitations to encode certain morphological properties of nouns or lack of certain types of modifiers. Accordingly, common nouns and their morphosyntactic behaviour have been viewed as the norm and the standard against which proper names have been evaluated. Two things have been neglected by this focus on common nouns and their properties as the starting point. First, the "defectiveness" of proper names can in many cases be explained by their special semantic and pragmatic status. Second, proper names respectively can have their own special morphosyntactic patterns not found with common nouns.

This paper aims at investigating proper names – mostly in the form of personal names, as will be discussed in the next section – in their own right, and not as defective as compared to common nouns. Since a full-scale study of the morphological and syntactic properties of names in the languages of the world is still a desideratum, this paper is limited to discussing illustrative examples without making any claims on how common or rare certain structures are in crosslinguistic comparison. The data presented are not meant to give a full typological overview. It rather focusses on presenting languages that have interesting and morphosyntactically rich marking for proper names. Examples are drawn from my ongoing research on the morphosyntax of personal names, for which I have checked the descriptions of about 250 languages so far (not all of which include information on names at all).

The structure of this paper is the following: In Section 2, a brief overview on previous work on linguistic differences between proper names and common nouns, as well as differences between different name classes, is given. Afterwards, such differences are illustrated aiming at including a set of languages as diverse as possible (both, genetically and areally). Section 3 focuses on the encoding of morphological categories such as gender, number, or case. Syntactical properties, mainly the combination with modifiers and appositional structures, are then tackled in Section 4. Finally, the findings are summarized and implications for further research are discussed in Section 5.

2 Not all names have been created equal

So far, either the everyday-term *name* has been used to refer to the central object of study of this paper or the more scientific one *proper name*. In this section, the terminology of the linguistic domain of names and nouns will be discussed (2.1). Afterwards, different types of names and possible grammatical differences between these are briefly presented (2.2).

2.1 Terminology

The primary distinction within the wordclass of nouns is often made between common nouns and proper nouns.[3] In the literature on names, the term *proper name* is used rather than *proper noun*. While proper noun is a lexical category and refers to syntactically simplex constructions, i.e. single words, a proper name is a functional category referring to the level of discourse. This means that a proper name is a linguistic entity that is used as a unique identifier in a given discourse context. Hence, proper names can be simplex or complex constructions; they can be established on the lexical level or ad hoc coinages.

This distinction is related to the one between a proper name and a *proprial lemma*, which as Van Langendonck (2007: 7–8) points out has been neglected in most of the onomastic and philosophical literature on names. Though in this case the relevant factor is only whether we are referring to a member of the lexicon (*proprial lemma*) or the function an element has in discourse (*proper name*). Since complex constructions, e.g. idiomatic expressions, are also member of the lexicon, the complexity of an item is not strictly relevant for this distinction.

In the following, I will also use the term proper name instead of proper noun, even though most of the examples contain proper nouns in the narrow sense of simplex, lexically established items. Since such items cannot be taken for granted to be found in all languages of the world, the level of usage – on which names are found universally – has to be the level investigated in cross-linguistic comparison, as pointed out by Schlücker and Ackermann (2017: 312).

---

[3] Whether proper nouns are indeed a member of the same wordclass as common nouns is a matter heavily debated and a major difference between the analyses of Anderson (2007) and Van Langendonck (2007). I will not go into any details on this issue here.

2.2 Name classes and differences between them

So far, the impression might have been given that names are a uniform class all sharing the same morphological and syntactic peculiarities. The actual situation is quite different, since names can be further classified according to the type of entity they are assigned to into names of persons (*anthroponyms*) – which in turn can be subclassified into given names, family names, nicknames and so forth – names of places (*toponyms*) – again showing a subdivision into names of cities, countries, mountains, forests etc. – names of historical events, institutions, works of art and many more (for a detailed classification of different types of names cf. Nübling et al. 2015: 101–106). In a given language, these different name types are unlikely to exhibit a uniform morphosyntactic behaviour.

Personal names are argued to be the most prototypical name type (Anderson 2003: 373; Van Langendonck and Van de Velde 2016: 33; Nübling et al. 2015: 104) and the one that is most likely to exhibit morphosyntactic differences from common nouns. Place names are the second (and only other) category one is most likely to encounter in the description of a lesser-known language (even though information on both types of names is regrettably limited in the large majority of grammars). Grammatical differences between those two name-categories are investigated by Stolz et al. (2017). The differences in the grammatical behaviour of place names they investigate can be broadly summarized as being of three types: distinct markers/allomorphy for certain grammatical categories (Chamorro articles, Kiribati prepositions), zero-coding of local relations (investigated in more detail in Stolz et al. 2014), and inability to serve as verbal argument (Classical Aztec). Stolz and Levkovych (this volume) present additional data on the (mostly) distinct grammatical behaviour between common nouns, personal names and place names in five Oceanic languages. Examples of the unlike behaviour of personal and place names are also discussed below (3.1) and a possible explanation is proposed. While trying to include data from place names whenever possible, the data presented in the following are heavily biased towards personal names.

3 Morphological marking

Despite their association with common nouns with respect to wordclass membership and/or typical syntactic function in an utterance, differences in the encoding of nominal categories have been long noted for nouns in a number of unrelated languages. The term "morphological" marking from the section heading is to be understood in a wide sense and not restricted to bound morphology. Depending on the language type, a formative of any of the categories investigated here might as well be expressed by a phonologically free form (cf. Bickel and Nichols' 2007: 172–193 discussion of inflectional morphology and the various ways it can be coded crosslinguistically).

In the following sections, the encoding of four morphosyntactic categories associated with nominals, namely definiteness (3.1), case (3.2), number (3.3), and gender (3.4) is surveyed. The fifth category associated with nominals – possessor-marking – will not be discussed here since examples of proper names with a possessor are highly unlikely to be found in reference grammars even if this construction is possible in the language. The findings are then summarized in Section 3.5.

3.1 Definiteness (and related concepts)

The nominal category usually referred to as definiteness not only encompasses proper definites in a semantic sense but can, depending on the language, also apply to the distinction between specific and non-specific items and related categories. Elsewhere (Handschuh 2017: 491), I have introduced the term *D-marking* – coined in parallel to the term D-element employed by Himmelmann (1997: 6–7) – in order to highlight the fact that a wider range of meanings are of interest. In this paper, this wide notion of definiteness is also applied, and the detailed semantics of the markers presented are not discussed. Proper names are per definition inherently definite (Van Langendonck 2007: 154) since they have a uniquely identifiable referent. Hence, they are also always specific and a careful distinction between these concepts does not seem to be necessary here. It has been argued (e.g. Van Langendonck 2007: 157), that due to their inherent definiteness, proper names require no overt marking through an article, explaining the absence of definite articles with names in many Indo-European languages, for instance in English. According to the (albeit small) sample in Handschuh (2017: 498–499), this pattern of overt marking of definiteness on common nouns but not on personal names appears to be an areal phenomenon of Western Europe and West Africa, while in other areas of the world definiteness is either or not overtly encoded for both types of nominals preferably.

While Handschuh (2017) only takes personal names as the prototypical name category into account, differences between the usage of articles with personal names and place names appear to be quite common. Already for the European languages that shun the usage of definite articles with personal names, place names show a more varied picture. With German country names, for instance, the article is possible but dis-preferred, cf. *(das) England* 'England', except for a few examples that require the article, e.g. *die Schweiz/*Schweiz* 'Switzerland'. Names of rivers on the other hand always use the article, cf. *die Donau/Donau* 'the Danube', *der Rhein/*Rhein* 'the Rhine'. D'hulst et al. (this volume) discuss the consistent co-appearance of river names with (preposed) definite articles in European languages and offer a striking semantically-based explanation. Yet, differences between personal names and place names can be found in other parts of the world as well (on the usage of definite articles with city names cf. Nübling et al. 2015: 207 on German and Caro Reina 2020 on Romance languages). Definite articles and proper names are also the topic of several contributions of this volume. Helmbrecht (this volume) offers a crosslinguistic survey of 40 languages that use an article with common nouns and investigates its presence or absence with different classes of names (outside of Europe, this is limited to personal names and place names). Salaberri (this volume) sketches the historical development of the occurrence D-marking on Basque proper names and common nouns and illustrates how the marker has grammaticalized into an element that is found on common nouns exclusively in the present language. Another historical analysis is provided by Caro Reina (this volume), who investigates the occurrence of definite articles with a variety of name classes in ten Romance languages.

In the following, I will present data from overt marking of definiteness with personal names. Examples of languages that have no such marking on proper names are excluded here (cf. Handschuh 2017 and Helmbrecht, this volume for such examples) since the aim of this paper is to demonstrate that names are not always as bare of morphosyntactic marking as is often suggested. The first language (Southeastern Tepehuan) employs identical forms with common nouns and personal names, the second language (Rapa Nui) uses distinct forms. Furthermore, both languages exhibit a clear distinction between personal names and place names, with the latter not being combined with articles – a fact for which another language (Dakaaka, related to Rapa Nui) offers an interesting explanation.

First, I discuss the data from Southeastern Tepehuan (Uto-Aztecan). They demonstrate that this language uses the article *gu* with common nouns (1) and personal names (2), but not with place names (3). These examples illustrate that not in all languages personal names are the category most prone to grammatically deviate from common nouns.

(1) *gu   chio'ñ* (Willett 1991: 206)
    ART man
    'the man'

(2) gu   Juan (Willett 1991: 80)
    ART   John
    'John'
(3) Susba'n-tam (Willett 1991: 80)
    frogs-place
    'Frog Town'

In the Oceanic language Rapa Nui a threefold distinction between common nouns marked with the article *te* (4), personal names marked with the article *a* (5) and zero coded place names (6) can be found.

(4) te   ika (Kieviet 2017: 239)
    ART.CN   fish
    'the fish'
(5) a   Hotu (Kieviet 2017:102)
    ART.PN   Hotu
    'Hotu'
(6) Rapa Nui (Kieviet 2017: 102)
    Rapa Nui
    'Rapa Nui'

A similar situation holds in a number of other Oceanic languages. Von Prince (2015: 136) argues that in Daakaka all place names (and other words referring to locations) are adverbs instead of nouns and have to be nominalized in order to serve as the argument on a verb or also as term of address (7). In Daakaka, nominalization of local adverbs is achieved by pre-posing the noun *vilye* 'place' (von Prince 2015: 112–114).

(7) O   vilye   Ambrym ... (von Prince 2015: 114)
    INTJ   place   Ambrym ...
    'Oh Ambrym'

This analysis of place names as adverbs has crucial implications for the debate on the wordclass-membership of names, since this would mean that the question has to be answered separately for distinct categories of names. Especially so, if this is not only a feature of Daakaka and related Oceanic languages but can be found in other language families as well. The data from Southeastern Tepehuan (1)–(3), that employs the article with common nouns and personal names but not with place names, demonstrate that similar patterns are found outside of Oceania. As Stolz et al. (2014, 2017) have shown, place names are particularly common to be zero-coded in the spatial relations of PLACE and/or GOAL, a situation that would be expected if indeed they are classified as adverbs (at least in some languages). The adverb status of place names would equally account for their inability to occur as the direct object of the verb 'to see' in Classical Aztec, which is described in Stolz et al. (2017: 140–142).

3.2 Case

While definiteness (and hence specificity) is inherent in proper names and thus arguably does not require overt marking, there is no a priori reason why one would expect names to behave any differently from common nouns for the category case, that marks the grammatical relation of an NP in the given utterance. Still, case marking is not always uniform for proper names and common nouns, as will be demonstrated in this section. Some very intriguing data on the distinct case marking of names and common nouns can be found in East-African languages. First, in two unrelated languages, Sinyar (usually, though not uncontroversially, classified as a Central Sudanic language) and Libido (a Cushitic language), names exhibit more distinct case forms than common nouns, contrary to the German tendency for personal names to have a very reduced case paradigm (cf. Ackermann 2018: 21). Another instance of a language with more distinct case forms for proper names than for common nouns can be found in Meriam of the Torres Strait Islands. The neighbouring, yet not genetically related, language Kalaw Lagaw Ya even has two distinct alignment systems for proper names (nominative-accusative) and common nouns (ergative-absolutive). In Kambaata, another Cushitic language, an instance of differential case marking is found in a rather specific context, namely in the *be-called* construction.

The case-marking system of Sinyar – and in particular the distinct case paradigms for proper names and common nouns – have been discussed in detail in Boyeldieu (2019). As can be seen when comparing the case forms marked on proper names (Table 1) and common nouns (Table 2), names distinguish five case forms (at least in singular) while common nouns only have three distinct forms. For common nouns, only the Adverbial (used for non-argument function) and the Nominative are distinguished from the so-called Absolute[4], which ap-

---

4 The *absolute* case is a label often employed in the grammatical description of so-called *marked nominative* languages (König 2006; Handschuh 2014). It is used to refer to the (usually)

pears in all other functions including citation form, object of a (transitive) verb and adnominal possessor. Names have specific forms for the two latter functions, namely the Accusative and Genitive. In addition, while the exponents of those case forms that both nominals have in common are not without overlap (two of the allomorphs of the Nominative Singular are shared), proper names and common nouns mainly use different forms for the cases marked on both.

Table 1: Case marking with proper names in Sinyar (Boyeldieu 2019: 479).

	(absolute)	Nominative	Genitive	Accusative	Adverbial
SG	unmarked	-n / (ˈ)-ñ / -lè[5]	-nà¹ / -nâ¹	-(y)àà	º-lèè[6]
PL		-ngè		-ngàá	º-ngèèr

Table 2: Case marking with common nouns in Sinyar (Boyeldieu 2019: 475).

	(absolute)	Nominative	Adverbial
SG	unmarked	-n / (ˈ)-ñ / -Ní / (ˈ)-Nĩ	-tí / (ˈ)-tĩ
PL		-sí / (ˈ)-sĩ	

Though supposedly unrelated (though as a language of potentially mixed origin, Sinyar might after all have a historically connection), Libido shows a similar picture. This Cushitic language also has more distinct case forms found on (the majority of) person names – while distinguishing more case forms than Sinyar in general – and, in addition, makes a distinction between different genders (for common nouns and person names) and inflection classes with names (Crass 2014).

Outside of Africa, two languages of the Torres Strait Islands exhibit interesting patterns in the case marking of common nouns and proper names. The lan-

---

least overtly-coded case form of a noun, that, in addition, carries a wide range of functions (the most typical being the citation form of a noun, and often the object of a transitive verb). What is special about these languages is that the function of subject of (transitive and intransitive) verb is not among the uses of the absolute case, but a special (more marked) form (the nominative) is used in this context (hence the name marked nominative).

5 The symbol ˈ marks a tonal downstep.
6 The symbol º marks that a preceding tonal downstep is neutralized by the respective case suffix.

guages Meriam (of the "Papuan" Trans-Fly family) and Kalaw Yagaw Ya (of the Australian Pama-Nyungan family) are not genetically related, yet it is worth noting that the two communities live in close proximity, are culturally very similar, and the two languages have a large set of shared vocabulary. Similar to Sinyar, Meriam has a special case form for the object of a transitive verb only for proper names. This is exemplified by the place name *Mer* in example (10) marked by the Accusative suffix *-i*. However, the general alignment system of Meriam is not of the nominative-accusative type (like in Sinyar), but ergative-absolutive. Thus, subjects of intransitive verbs are in the Absolutive case (that has no overt marker) for both proper names (8) and common nouns (11). The same form is also used with objects of transitive verbs for common nouns (13). Transitive subjects on the other hand are in the Ergative case for both proper names (9) and common nouns (12) that is encoded through the suffix *-ide* (or one of its allomorphs like *-ut* in 13).

(8) **Mesnare** *ta-bakyamu-da*
 Mesnare.ABS SG.S-go-PFV.SG
 'Mesnare came.' (Piper 1989: 33)
(9) **Lez-ide** *tabara imus itw-i.*
 Lez-ERG 3SG.GEN beard.ABS shave-PFV
 'Les shaved his (own) beard.' (Piper 1989: 69)
(10) *Ka* **Mer-i** *dásmer-i mi ebur-ge.*
 1SG.A Mer-ACC see-PFV sky animal-LOC
 'I saw Murray Island from the plane.' (Piper 1989: 33)
(11) *Able aw* **le** *áb-i.*
 ART big person.ABS fall-PFV
 'The old man fell.' (Piper 1989: 34)
(12) **Lamar-ide** *kári na-ge-li.*
 spirit-ERG 1SG.O 1/2SG.O-frighten-PRS.IPFV
 'A ghost has been frightening me.' (Piper 1989: 32)
(13) *Able kebi lel-ut aw* **le** *eparsi-da.*
 ART little person-ERG big person.ABS hit-PFV.SG
 'The youth punched the old man.' (Piper 1989: 34)

In Kalaw Lagaw Ya, common nouns and proper names (the description speaks of proper nouns; however, the examples only contain personal names) even exhibit completely opposite alignment systems. While common nouns are of the ergative-absolutive type as in Meriam, proper names have a nominative-accusative system. As can be seen in (14), the Absolutive case (without overt marking) of the common nouns *garkaz* 'man' and *burum* 'pig' occurs with intransitive subjects. The same form occurs with the transitive object in (15), while the transitive subject is marked by the Ergative suffix *-in* (the change in the noun stem from *garkaz* to *garkoez* is phonologically conditioned). For proper names, the form of the intransitive subject – devoid of overt marking – illustrated in (16) coincides with the form of the subject of the transitive verb in (17), while the object in this example receives the Accusative suffix *-n*.

(14) **Garkaz   /   burum**   *uzariz.*
man.ABS   pig.ABS   went_away
'The man/pig went away.' (Comrie 1981: 8)

(15) **Garkoez-in   burum**   *mathaman.*
man-ERG   pig.ABS   hit
'The man hit the pig.' (Comrie 1981: 8)

(16) **Kala   /   Gibuma**   *uzariz.*
Kala.NOM   Gibuma.NOM   went_away
'Kala/Gibuma went away.' (Comrie 1981: 8)

(17) **Kala   Gibuma-n**   *mathaman.*
Kala.NOM   Gibuma-ACC   hit
'Kala hit Gibuma.' (Comrie 1981: 8)

Back in Africa, a more minor difference in the case marking patterns of common nouns and proper names can be found. Kambaata, a Cushitic language (closely related to Libido), employs the same case paradigms for proper names and common nouns, yet the usage of the cases shows a slight variation between different types of nominals. In the construction introducing the name/noun by which someone (or something/somewhere) is being referred to – the *be-called* construction – a striking pattern emerges. While a personal name in this construction is used in the Vocative (18), if a common noun is used in reference to a human being, this noun is in the Nominative instead (19), even though Vocative forms exist for all types of nominals in this language. And finally, when giving the name of a place or the word that designates an inanimate object, the form that is used is the Accusative (20).

(18) *Ánn-u-'[i]*   **Wotáng-o**   *y-am-am-áno*
father-NOM-1SG.POSS   Wotang-VOC   say-PASS-PASS-3.IPFV
'My father is called (lit. said) Wotango.' (Yvonne Treis p.c.)

(19) *Rosisáanch-u-se*   *ti*   **qoxár-at**   *y-ée-se*
teacher-NOM-3.POSS   DEM.NOM   clever-NOM   say-3.PFV-3.OBJ

'The teacher called her a genius.' (Yvonne Treis p.c.)

(20) *Kú=b-u*   ***Duuraam-íta***   *y-am-am-áno*
DEM.NOM=PLC-NOM   Duuraame-ACC   say-PASS-PASS-3.IPFV
'This place is called Duuraame.' (Yvonne Treis p.c.)

This instance of *differential case marking* between personal names, common nouns and place names demonstrated that constructions of naming and/or being-called offer another interesting research field for the crosslinguistic study of names.

3.3 Number

Despite the general observation that proper names usually only occur in singular "[b]ecause of their fixed denotation" (Van Langendonck 2007: 159), number marking of names is a more complex issue than one would suspect. First, distinct name classes differ with respect to whether they refer to a singular entity or to a group. This is most obvious with personal names, for which given names and family names exemplify the two options. Especially, family names are of interest here, since they can be both used for singular and plural reference. Since for this phenomenon hardly any data outside of European languages can be found, I will not go into much detail here. At the end of this section, I will briefly contrast data from German and French, which exhibit different degrees of morphological impoverishment for the marking of plurality on family names (not surprisingly, I have not come across any language exhibiting a richer morphological system than with common nouns in this context). Second, with the associative plural, a crosslinguistically very common type of plural marking exists for which personal names are central.

The construction of associative plural is quite well investigated and described crosslinguistically (among others: Moravcsik 2003; Daniel and Moravcsik 2013; Mauri and Sansò 2019). Hence, this section will be kept rather brief. The associative plural contrasts with the regular (or additive) plural in that it does not refer to several instances of the pluralized entity but refers to a group that is associated with the pluralized entity. This pluralized entity is referred to as the *focal referent* (Moravcsik 2003: 471), and the group associated with it most typically is the family of said entity though the exact semantics of the construction are not fixed and usually depend on the context. Associative plurals can either have marking that is exclusively used for this type of plurality (as exemplified by Atong below), or more general plural markers are used that can either have an additive or associative meaning depending on context (as is the case in Turkish).

The following example from Atong (Tibeto-Burman) illustrate the use and meaning of the associative marker =*para* (21). As the name suggests, this marker is used to encode the associative meaning only.

(21) aŋa **letit**=***para***=məŋ   nok=ci   saʔ-ni
     1SG   Latith=&co=GEN   house=LOC   eat-FUT
     'I will eat at the house of Latith and company.' (van Breugel 2008: 116)

Associative plurals are most commonly found with personal names and kinship terms and in some languages even limited to them (Mauri and Sansò 2019: 603). In Atong, the associative plural only occurs with nouns referring to humans, as well as anthropomorphized animals in stories (van Breugel 2008: 116). As has been mentioned above, apart from the extension of the associative plural within the nominal domain variation between languages is also found with the range of meanings the marker encoding associative plurality has. The first of two general patters has been illustrated by the specialized marker in Atong. The second option is for "regular" plural marking to have an associative interpretation, when the semantics of the noun it is used with do not allow for an additive plural interpretation. This is the case in Turkish, as is illustrated by the plural of the personal name in (22) and common noun in (23).

(22) **Ahmet-ler**   gel-me-di.
     Ahmet-PL   come-NEG-PST
     'Ahmet and his spouse/family/friends did not come.' (own knowledge)
(23) **Erkek-ler** gel-me-di.
     man-PL   come-NEG-PST
     'The men did not come.' (own knowledge)

As noted above, for family names a regular (additive) plural sense – referring to several members of the same family – exists in addition to the possibility to use the family name alone, when referring to a single person (a practice found, for instance, with citations in academic writing). The plural marking of additive plurals on family names is not studied for languages outside of Europe and information on this topic is very unlikely to be included in descriptive grammars of lesser-known languages (if they even have indigenous family names or equivalent structures). For Europe, the well-established picture of names exhibiting less variation than common nouns can also be found in this domain. Either, no overt marking of plurality is possible, as in French (cf. *les Chirac*) and other Romance languages (Mańczak 1995: 427), or the inventory of forms is strongly reduced as compared to common nouns. The latter is the case in German, that only allows for the default *s*-plural with family names even in cases where a homophonous common noun with a different plural-allomorph exists, e.g. *der Mann/die Männ-er* 'the man/men' vs. *Thomas Mann/die Mann-s* 'Thomas Mann/the Mann family' (Ackermann 2018: 23–24).

3.4 Gender

Since the category of grammatical gender can always be traced to a sematic core in which assignment is based on some property, such as biological sex or animacy, of the referent (Corbett 1991: 8), and since most often this core of the system is the distinction between male and female (human beings), the assignment of personal given-names to a specific gender – if the language has this category – should be straightforward. And indeed, as I have discussed in Handschuh (2019: 543–547), this assumption is true for many languages. However, not in all languages personal names are simply assigned to the same class as nouns referring to humans (of the same sex). The main exceptions already discussed in Handschuh (2019) will be briefly summarized here. These exceptions are the exclusion of names from the gender system of a language (Eton, Yimas) and the existence of (a) special gender category/categories for names (Teop, Bukiyip – since the latter language has not been addressed in the previous paper, it will be discussed in more detail here).

For some languages, it has been argued that (personal) names are not part of the gender system altogether. The most prominent example is the Bantu language Eton (Van de Velde 2006, 2008), but a parallel claim has also been made for Yimas (Foley 1991: 170). The different genders in Eton are referred to as noun classes (as is common practice in Bantu linguistics, henceforward NC), which also encode singular/plural-distinctions, and which are simply labelled by numbers. Personal names in Eton are traditionally considered to be members of NC 1a, a sub-gender of NC 1, to which words referring to persons (in the singular) typically belong. Van de Velde (2006) argues that the morphosyntactic behaviour of class 1a nouns is very different from nouns belonging to other classes for two main reasons, and that thus they are genderless. The first reason is the morphological structure of the noun itself. While NC 1 is marked by the prefix ǹ-on the noun, cf. *m-ìnŋgá* 'woman'[7], which corresponds to the NC 2 prefix *b(è)-* when in the plural, cf. *b-ìnŋgá* 'women' (Van de Velde 2008: 84–85). This overt encoding of noun-class membership is not found with personal names (and other members of NC 1a, like kinship terms). Personal names can be derived from common nouns via the suffix *-ɔ́* (accompanied by a tonal change), as can be seen when comparing the common noun in example (24) with the resulting personal name in (25). While the common noun has an overt marker of its noun class via prefix (*è-*), this prefix is incorporated into the stem of the proper name, which does not belong to NC 5 and, unlike the common noun it is based on, has no plural reference. Since plural marking is not viable through noun-class marking, a different plural strategy is employed by personal names, as is demonstrated with the associative plural marker *bɔ̀* (26).

(24) *è-nwăn* 'birds (NC5)' (Van de Velde 2008: 96)
(25) *ènwànɔ́* 'Birds (personal name)' (Van de Velde 2008: 96)
(26) *bɔ̀ ènwànɔ́* 'Birds and his friends' (Van de Velde 2008: 97)

While personal names (and other NC 1a nouns) differ from the remaining members of NC 1 (referred to as NC 1b) with respect to the overt encoding of noun-class membership via prefix, both types of nouns trigger the same agreement markers (e.g. on the verb). Apart from their shared semantics of having human referents, the usage of identical agreement markers is the main reasoning of assigning personal names to NC 1. Van de Velde (2006) argues that agreement markers can either be assigned based on the noun-class membership of the agreement trigger, but there is also a rule of default-agreement. The set of default agreement markers is the one also used with NC 1 nouns, and default agreement is triggered if no controller gender is available (this is the case for instance with interjections). This is exactly the scenario he claims is the case with NC 1a nouns as well.

The situation in Teop (Oceanic) is completely different, in this language, personal names are clearly assigned to a gender, and moreover they are the core members of a distinct gender (or sub-gender). Teop is one of the very few Austronesian languages that have developed a fully-fledged gender system. This system is historically related to the wide-spread division between the common and personal (or proprial) article, that we have already encountered in Rapa Nui (examples (4) and (5) above). Yet, the Teop gender system is not restricted to distinct forms of the article but is mirrored in similar markers on other nominal

---

[7] The initial nasal of the prefix is underspecified for place of articulation and changes according to the first segment of the noun stem it attaches to.

modifiers (like adjectives) that agree with their head noun in gender (and, also in number). Mosel and Briggs (2000: 322–323) distinguish two basic genders – I and II – based on the forms used on agreement targets, while they introduce a subdivision of the first gender into I-E and I-A based on (and labelled after) the form of the article co-occurring with the head noun in the singular. The nouns of gender I-A are personal names, kinship terms, nouns denoting pets and persons with a particular status in the community (Mosel and Briggs 2000: 334). Gender I-E, on the other hand, is the class that all other words referring to humans and animals, as well as a number of inanimate objects related to daily life (e.g. food items, utensils, landmarks) belong to; it is considered the default gender by Mosel and Briggs (2000: 336–338). The last gender – number II – comprises a diverse list of mostly inanimate items, and also most mass and abstract nouns (Mosel and Briggs 2000: 338–340). The following examples show nouns of gender I-E (27), I-A (28), and II (29) with their respective article.

(27) e *Kakato* 'Kakato' (male name) (Mosel and Briggs 2000: 334)
(28) a *otei* 'the man' (Mosel and Briggs 2000: 336)
(29) o *kasuana* 'sand' (Mosel and Briggs 2000: 339)

Another language, for which it is claimed that proper names are members of a distinct gender class, is Bukiyip from the Torricelli family of Papua New Guinea (Conrad and Wogiga 1991). Moreover, personal names and place names are both assigned to a class of their own in this language. Conrad and Wogiga (1991) describe 18 classes distinguished through (mostly) distinct suffixes for the singular and plural of a noun as well as distinct markers on agreement targets such as adjectives and verbs.[8] The classes are mostly semantically arbitrary, though transparent assignment is found for female persons (Class 4), male persons (Class 7), persons of unspecified or mixed gender (Class 8), as well as personal names (Class 17) and place (Class 18) names (Conrad and Wogiga 1991: 8, 10). While for all other genders, the paradigm lists a complete set of six markers: nominal and adjectival suffixes for both singular and plural forms, as well as verbal prefixes in the singular and the plural, the paradigms for personal and

---

8 On which basis the distinct noun classes have been established, is not discussed by the authors of the grammar. As will be demonstrated below, despite being analysed as separate gender classes, neither personal nor place names have a separate set of markers not occurring with nouns from other genders. In the introduction, Conrad and Wogiga (1991: 1) mention that they have mainly followed the analysis of the gender system as found in the earlier description by Fortune (1942), a work I have not been able to consult so far.

place names have some gaps. For personal names (Class 17) no nominal suffixes are found. For adjectival and verbal agreement, only singular forms are listed that correspond to the respective forms of female (Class 4) and male (Class 7) persons. For place names (Class 18), the noun itself and an agreeing adjective have no suffix in the singular, but a suffix for the plural is listed, as well as verbal agreement-prefixes in singular and plural. In all cases, the affixes are homophonous with the ones in Class 15. This is, however, not as one could expect the same class the noun 'village' is found in, which belongs to Class 2 respectively. An overview of affixes from the classes relevant here is given in Table 3.

Table 3: Noun classes and their markers in Bukiyip (based on Conrad and Wogiga 1991: 10).

Noun class	Gloss of example	Exampel singular/plural	Noun suffix		Adjective suffix		Verb prefix	
			SG	PL	SG	PL	SG	PL
2	'village'	wabél; walúb	-bél	-lúb	-bili	-lúbi	bl-	bl-
4	'woman'	élmatok; élmagou	-k	-ou/-eb	-kwi	-wali	kw-	w-
7	'man'	élman; élmom	-n/-nú	-m	-nali	-mi	n-	h-
8	'child'	batawiny; batawich	-ny/-l	-ch/-has	-nyi/-li	-chi	ny-/l-	ch-
15	'small pig'	buligún	-gún	-gún	-gúni	-gúni	gn-	gn-
17	personal name		-	-	-nali/-kwi	-	n-/kw-	-
18	place name		-	-gún	-	-gúni	gn-	gn-

While the conclusions drawn by Van de Velde (2006) and Conrad and Wogiga (1991) on the status of names with respect to the gender system are diametrical, the data from Eton and Bukiyip is actually quite similar. Both languages generally have overt markers of noun class membership found on the noun itself, that is however absent from proper names. Agreement on the other hand is found to occur with names, while the markers are the same used for some other nouns, the ones of NC 1b (or default agreement if following Van de Velde's (2006) proposal) in Eton, and Class 4 (female personal names), Class 7 (male personal names), or Class 15 (place names) in Bukiyip. One minor difference is that the Bukiyip default gender is yet another gender, namely Class 8 (the one that persons with unspecified sex are found in). Nevertheless, both languages illustrate that names have properties unlike other nominals with respect to the encoding of gender – whichever conclusions one wants to draw from this – and this topic merits further studies.

3.5 Summary

This section has illustrated that special behaviour of proper names can be found in all four nominal categories investigated here and in languages of various families. While the often-noted tendency of names to allow for less variation and to avoid overt marking that changes the phonological shape of the name (referred to as "onymic schema constancy" by Nübling 2017) is present in a number of languages, the opposite can also be found. This is perhaps most impressively demonstrated by the wider range of case forms of proper names in Sinyar, Libido, and Meriam (3.2). Also, the claim that personal names are the category most likely to show distinct behaviour from common nouns has been challenged by the absence of (definite) articles with place names but not personal names in Southeastern Tepehuan and numerous Oceanic languages (3.1). A possible explanation of this has been provided based on the analysis of von Prince (2015) of Daakaka place names as adverbs and not nominals.

4 Syntax

For the syntactical behaviour of proper names, the tendency to highlight their defectiveness as compared to common nouns is especially common. Dixon (2010: 108) states that "[a] proper name as the head of an NP is likely to have far fewer – if any – possibilities for modification, when compared to a common noun as NP head". He further lists their possible inability to be modified by adjectives, relative clauses, or demonstratives (Dixon 2010: 108). While modification of proper names is restricted in some way, a syntactic context that names are primarily found in is the appositional constructions, more precisely what Van Langendonck (2007: 125–138) refers to as *close apposition* and even list as the defining feature for proper-namehood.

In the following, the behaviour of proper names in three syntactical contexts is surveyed. First, appositional constructions are discussed (4.1). Afterwards, the potential to combine with modifiers of the noun phrase as well as possible alternative constructions used with names are taken into consideration (4.2). And finally, some interesting data on the use of coordinators are presented (4.3), as well as a brief summary of the section (4.4).

4.1 Apposition

As noted above, Van Langendonck (2007) takes the ability to appear in close apposition as a defining criterion for proper-namehood. He illustrates this and the distinction between close and loose apposition (Van Langendonck 2007: 126–128) for three Indo-European languages (English, Dutch and French). Important criteria for the distinction are the possibility to use definite articles with either member of the appositional construction, their respective ordering, as well as the possibility to interchange the members and to delete either member of the construction. Such detailed syntactic analyses of appositional constructions are not found for most lesser-known languages, which are of central interest for the present study. Since a distinction between close and loose apposition is not possible based on the data analysed here, I will look at appositional constructions in general and only try to distinguish between the two types if the data allow for it. The working definition used here for apposition is that of two (or possibly more) distinct and co-occurring noun phrases that share the same referent and the same syntactical function in a clause.

The following examples from Rapa Nui (Kieviet 2017) illustrate the use of proper names as the first (30) and second (31) element in an appositional construction. In the first example, the personal name *Reŋa Roiti* is clarified to belong to a girl through the appositional noun *poki tamahahine*. This example appears to be a case of loose apposition based on the insertion of a comma after the name supposedly marking an intonation break, which often marks this construction and is never found in close apposition (Van Langendonck 2007: 126). In the second example, it is more probable that we are dealing with close apposition. Here the common noun *'ariki* 'king' precedes the personal name *Hotu Matu'a*.[9]

(30) *He poreko ko Reŋa Roiti, poki tamahahine*
 ASP born PROM Renga Roiti child female
 'Renga Roiti, a girl, was born.' (Kieviet 2017: 272)
(31) *He oho mai era te 'ariki ko Hotu Matu'a he rarama era*
 ASP go hither DIST ART.CN king PROM Hotu Matu'a ASP inspect DIST
 'King Hotu Matu'a came and examined it.' (Kieviet 2017: 273)

---

[9] This example may be controversial, as one reviewer pointed out, since the noun *'ariki* 'king' could also be analysed as a title that is part of the name itself. Since this example is analysed as an appositive structure in the respective grammar, I will present it as such here. Additionally, Van Langendonck (2007: 127) discusses "institutionalized titles such as *Queen Elizabeth*" as a subtype of close appositional constructions. However, with reference to Quirk et al. (1972: 621), he also notes that there is a gradual shift from close apposition to constructions that include full titles.

The construction in (31) is the most common one with proper names (Kieviet 2017: 512), however, as example (32) illustrates it is also possible that the name precedes the noun in what still appears to be close apposition.

(32) *He turu   a   Rovi, he   taŋata hāpa'o i   te   poki 'a Hotu*
     NTR go_down ART.PN Rovi PRED man   care_for ACC ART.CN child of Hotu
     'ariki
     king
     'Rovi came down, the man who took care of the child of king Hotu.'
     (Kieviet 2017: 274)

According to Kieviet (2017: 272), appositive structures are used instead of relative clauses – that are always restrictive in Rapa Nui – when providing information on a nominal that already has a unique referent. This, of course, includes all proper names like the place name *Rano Raraku* in the following example (33).

(33) *He   oho   ia   a   Vakaiaheva   ki   Rano Raraku,*
     ASP   go   then   ART.PN   Vakaiaheva   to   Rano Raraku
     kona ['i   ira   te   kape   e   noho   era].
     place at   ANA   ART.CN   boss   IPFV stay   DIST
     'Vakaiaheva went to Rano Raraku, the place where the boss lived.' (Kieviet 2017: 272)

In Maleu, an Oceanic language of the West New Britain Province of Papua New Guinea, appositional NPs are also very common. The following examples illustrate the occurrence of the personal name *Kamo* (34) and the place name *Gusei* (35) as the second member of an appositional construction. Both are preceded by a common noun that specifies which type of entity we are dealing with. Another common structure is for a personal name to be pre-ceded by a kinship term (36), that in this case shares the semantic property of unique identifiability with the name. The only instance in which a personal name is found as the first member of an appositive construction is in combination with a reflexive element like *tauaŋa* (37).

(34) *na-vola tako Kamo tako*
     CN-man this Kamo this
     'this man Kamo' (Haywood 1996: 163)

(35) *na-ko   Gusei ai-a*
CN-water Gusei leg-3SG
'the river Gusei's leg' (Haywood 1996: 165)

(36) *tam-e   Topi*
father-3SG Topi
'his father Topi' (Haywood 1996: 167)

(37) *Kamo ie   tauaŋa*
Kamo 3SG REFL.3SG
'Kamo himself' (Haywood 1996: 167)

In both languages, there is a clear tendency for proper names to appear as the second element in an appositive construction. More detailed studies of possible deviations from this pattern, especially with respect to Van Langendonck's (2007) discussion of close and loose apposition, for these and other languages would certainly be very welcome. Only then, could we make any hard claims about whether this ordering preference is indeed a universal tendency (such as: more general information precedes more specific information), or whether language specific parameters (such as basic word order) have some influence on appositional structures. It should be noted that the English data Van Langendonck discusses show the same preference, even though English is an SVO language and the Oceanic languages are verb-initial.

4.2 NP modifiers

As stated in Dixon's (2010: 108) quote above, proper names are known to have a restricted combinatory potential with modifiers associated with the noun phrase. Van Langendonck notes that, more specifically, the use of restrictive modifiers is not possible with proper names, since those "modifiers limit the extension of a given NP" (2007: 143), and since named entities are per definition already uniquely identifiable such a restriction of possible referents is simply superfluous. Of course, this restriction does not hold for the de-onymic use of proprial lemmata (hence Van Langendonck's careful distinction between the lexical and syntactic category).

Most likely, modification of proper names will be the exception and not the rule in a language. A very detailed overview of the possible modifiers to appear in the proper noun phrase is given by Kieviet (2017) for Rapa Nui. First, he notes that "most proper noun phrases consist only of a proper noun preceded [...] by the proper article *a*" (Kieviet 2017: 274). He then continues to describe four slots for pre-nominal elements and six slots for post-nominal elements that can optionally appear with a proper name (though highly unlikely at the same time). These are, of course, far less possibilities than can be found in the common noun phrase, for which he lists seven positions before and ten positions after the head noun (Kieviet 2017: 229–230), but this demonstrated that proper names can be syntactically more interesting than the broad claim by Dixon (2010) suggests.

Above (4.1), Kieviet's (2017) claim that appositional constructions are an alternative strategy for relative clauses with proper names in Rapa Nui has already been mentioned. Another example of the difficulties to modify names with relative clauses is illustrated by Hmong Njua (Harrienhausen 1990). The first example illustrates the modification of a common noun *mi ntai* 'girl' with a relative clause. It is the S-argument of the main-clause verb and modified with a pre-posed classifier (*tug*), a postposed demonstrative (*nuav*), and the relative clause, that follows after the main verb and is marked with the identical classifier as its head noun (38). If a personal name on the other hand is to be accompanied by a relative clause, the construction looks a bit different. In this case the head NP does only consist of the demonstrative (that is occupying the formal head position now) and the preceding classifier, while the proper name is only inserted after the main-clause verb as a kind of afterthought-topic (39). A more literal translation of the example might be 'This is the one whom I met yesterday, Peter'.

(38) tug mi ntai nuav yog tug kws kuv tau ntsiv tom tas les
     CLF girl   DEM be CLF REL 1SG PST meet at   beach
     'This is the girl whom I met at the beach.' (org.: 'Dies ist das Mädchen, das ich am Strand getroffen habe'; Harrienhausen 1990: 142)

(39) tug nuav yog Peter tug kws kuv tau ntsiv naag mo
     CLF DEM be Peter CLF REL 1SG PST meet yesterday
     'This is Peter whom I met yesterday.' (org.: 'Dies ist (der) Peter, den ich gestern getroffen habe'; Harrienhausen 1990: 142)

Unfortunately, no context is provided for the above examples. From the original translation into German, the first example appears to have a restrictive meaning (unlike in English, punctuation does not clarify the reading of the relative clause). The second example could be either a restrictive relative clause with the name *Peter* used as a common noun ('a person named Peter'), in which case the definite article *der* would preferably be used, or a non-restrictive relative clause with *Peter* functioning as a proper name, in which case Standard German would not use the definite article, which would however be used in many colloquial varieties. Despite these uncertainties, the example illustrates that relativizing names (either those actually functioning as proper names or proprial lemmata in common noun usage) is no viable option in this language and a demonstrative is used as the head of the relative clause instead with the name postposed as an afterthought-topic-like construction.

A different alternative strategy to avoid names (even in common noun usage) with restrictive modifiers is found in Geba Karen (Shee 2008). In this Sino-Tibetan language, proper names cannot be directly combined with numerals. While common nouns occur in a construction consisting of the noun itself followed by the numeral, which is then followed by a classifier (40), a parallel construction with a personal name is ungrammatical (41). In a context where a proprial lemma has to be combined with a numeral – this is of course only possible when the name is used as an appellative with the meaning 'several persons bearing this name' – an alternative construction can be used (42). In this alternative construction the verb ʔɔ́, glossed as 'have', is placed after the proper name and before the numeral. The range of uses found in Shee's (2008) grammar suggest that this verb functions as a copular or existential element in Geba Karen.[10] A more literal translation of example (42) would be 'We have two, who are (called) Maung, at school'.

(40) mō   θó   bwὲ
 mother three CLF
 'three mothers' (Shee 2008: 39)
(41) *maòŋ dʒì bwὲ
 Maung two CLF
 Intended: 'two Maungs' (Shee 2008: 40)
(42) maòŋ ʔɔ́   dʒì bwὲ dó tʃaúŋ bú nò
 Maung have two CLF at school in FIN
 'There are two Maungs at school.' (Shee 2008: 41)

As the examples above have shown, restrictions on names in the combination with nominal modifiers exist in many unrelated languages. These restrictions may even be on the lexical level, excluding proprial lemmata from certain constructions even in de-onymic usage. However, the data above have also demon-

---

10 The usage of the verb ʔɔ́ 'have' as a copula is illustrated by the following example:
 a) dὲ   gò   ʔɔ́
   thing hot   have
   'It's very hot.' (Shee 2008: 123)

strated that these restrictions can be circumvented by using alternative – arguably more marked – constructions.

4.3 Coordination

The last syntactic context that will be investigated here is coordination. Since items that can be coordinated must be of the same syntactic category, this construction might be able to shed some light on the syntactical status of proper names in a given language. I have not encountered any detailed discussion of the coordinative possibilities of proper names in a lesser-known language so far. A different interesting phenomenon for the coordination of names will be presented here instead.

In some Oceanic languages – namely in Nemi, Nēlēmwa, Drehu, and in languages of the Central Pacific subgroup of Eastern Polynesian – a different coordinator is used with proper names than with common nouns according to Moyse-Faurie and Lynch (2004: 453). In Nemi (New Caledonia), proper names use the coordinator *ma* (43). The usage of this marker extends beyond its core function to coordinate proper names and is also found with definite common nouns referring to humans (Moyse-Faurie and Lynch 2004: 454). Otherwise, common nouns either employ the "tight" coordinator *men* (44), which is used with naturally co-occurring pairs of items (Moyse-Faurie and Lynch 2004: 450), or the "loose" coordinator *o* (45).

(43) *Kaavo ma   Hixe*
 Kaavo and.PN Hixe
 'Kaavo and Hixe' (Ozanne-Rivierre 1979 cited after Moyse-Faurie and Lynch 2004: 454)
(44) *kut men   daan*
 rain and.tight wind
 'rain and wind' (Ozanne-Rivierre 1979 cited after Moyse-Faurie and Lynch 2004: 451)
(45) *ngeli kuuk o   ngeli hyo*
 ART   yams and.loose ART   taro
 'the yams and the taros' (Ozanne-Rivierre 1979 cited after Moyse-Faurie and Lynch 2004: 451)

The historical origins of the distinct markers of coordination would be of great interest for these languages, a topic unfortunately not addressed in the cited paper. A well-known grammaticalization path for coordinators is of course the comitative(-instrumental). It seems plausible that a specialized personal name coordinator could have a comitative source. Based on the present information, this is however a mere speculation.

4.4 Summary

The previous sections have illustrated that the syntactic environment in which proper names appear are an understudied field. Appositive constructions have not been on the agenda of linguistics since the structuralist framework has been mostly abandoned, as noted by Van Langendonck (2007: 125). Yet, they provide an environment in which proper names appear to thrive in language after language, and the study of the syntax of names requires their in-depth investigation in lesser-known languages. As has been noted before, the potential of modification is often limited for proper names. However, one might also look at this from a different perspective and say that the need for modification is limited with proper names, since modifiers help to clarify the reference of a noun phrase, a function simply unnecessary for names that are already uniquely identifiable. Another aspect should be taken into consideration. The tendency of proper names to appear as the second element of an appositional construction has been described above. In the languages investigated here, this position is often analysed itself as a modifier position in the noun phrase, while the first element of the apposition is considered to be the head of the phrase. It should not come as a surprise that elements that preferably appear in modifier position are seldom modified themselves, since this would lead to a nested syntactic structure of the phrase and make parsing more difficult.

5 Conclusion

Proper names exhibit various grammatical differences from common nouns in the languages of the world. We are far from having a complete understanding how these differences are manifested. The view that names are an un-interesting, since defective, class of nouns is far too simplistic, and in many cases not true, cf. for instance the more elaborate case systems of proper names as compared to common nouns in Sinyar, Libido, and Meriam (3.2). Based on their special semantics of unique identifiability, as well as their special communicative functions (a topic that has not been treated here, since it is beyond the scope of this paper), they are used to different ends than common nouns. Differences in the syntactic environment they appear in are a logical consequence of this.

One question that has arisen from the discussion of modification of proper names is on which level possible restrictions are found: the syntactical one or the lexical one? Van Langendonck (2007) carefully distinguishes between proper names on the one hand and proprial lemmata on the other. While it seems plausible that the syntactic restrictions should be based on the syntactical function an element has in a specific environment, the inability of names to be directly combined with a numeral in Geba Karen, even if clearly used in appellative function, suggest that the restriction is based on the lexical category. This topic could not be investigated in full depth here, further research in this direction will certainly broaden our understanding of how and why proper names differ grammatically from common nouns.

The paper has also demonstrated that proper names do not necessarily exist as a monolithic category in the languages of the world. The two best-studied types of names – personal names and place names – can behave in a way that warrants assigning them to different parts-of-speech, as is the case in Daakaka. The more we investigate the grammar of names in detail, the more examples like this we are bound to encounter. Hopefully, the illustrative examples presented here can convince even more linguists to dig deeper into the grammar of names in the language(s) of their expertise.

Acknowledgements

I would like to thank the editors of this volume, Javier Caro Reina and Johannes Helmbrecht, as well as an anonymous reviewer for their helpful comments on this paper. Most of the data discussed has been presented at the 2019 DGfS-Workshop that this collection of papers originates from. In addition, I have read a paper containing the data investigated in Section 4 at the *Vielfaltslinguistik* conference held in Cologne in March 2019. The audiences on both occasions have given valuable comments. Special thanks are due to Kilu von Prince for drawing my attention to her analysis of Daakaka place names as adverbs. Nevertheless, the full responsibility for the contents of this paper, including all errors and omissions, remains with me.

Abbreviations

1, 2, 3	first, second, third person
A	most agentive argument of transitive verb
ABS	absolutive case
ACC	accusative
ANA	general anapher
ART	article
ASP	aspect
CLF	classifier
CN	common noun
DEM	demonstrative
DIST	distal
ERG	ergative case
FIN	final particle
FUT	future tense
GEN	genitive
INTJ	interjection
IPFV	imperfective
LOC	locative
NC	noun class
NEG	negation
NOM	nominative
NP	noun phrase
NTR	neutral aspect
O	least agentive argument of transitive verb
OBJ	object
org.	original translation
PASS	passive
PFV	perfective
PL	plural
PLC	place
PN	proper name
POSS	possessor
PRED	predicate marker
PROM	prominence marker
PRS	present tense
PST	past tense
REFL	reflexive

REL	relative clause
S	sole argument of intransitive verb
SG	singular
VOC	vocative
&co	associative marker

References

Ackermann, Tanja. 2018. *Grammatik der Namen im Wandel: Diachrone Morphosyntax der Personennamen im Deutschen* (Studia Linguistica Germanica 134). Berlin & Boston: De Gruyter.

Ackermann, Tanja & Barbara Schlücker (eds.). 2017. The morphosyntax of proper names. [Special Issue]. *Folia Linguistica* 51(2).

Allerton, D.J. 1987. The linguistic and sociolinguistic status of proper names: What are they, and who do they belong to? *Journal of Pragmatics* 11(1). 61–92.

Anderson, John. 2003. On the structure of names. *Folia Linguistica* 37(3–4). 347–398.

Anderson, John M. 2007. *The grammar of names*. Oxford: Oxford University Press.

Bickel, Balthasar & Johanna Nichols. 2007. Inflectional morphology. In Timothy Shopen (ed.), *Language typology and syntactic description. Volume III: Grammatical categories and the lexicon*, 169–240. 2nd edn. Cambridge: Cambridge University Press.

Boyeldieu, Pascal. 2019. Proper names and case markers in Sinyar (Chad/Sudan). In Antje Dammel & Corinna Handschuh (eds.), *Grammar of names*. [Special issue]. *Language Typology and Universals* 72(4). 467–503.

Caro Reina, Javier. 2020. The definite article with place names in Romance languages. In Nataliya Levkovych & Julia Nintemann (eds.), *Aspects of the grammar of names: Empirical case studies and theoretical topics* (LINCOM Studies in Language Typology 33), 25–51. München: LINCOM.

Caro Reina, Javier. this volume. The definite article with personal names in Romance languages.

Comrie, Bernard. 1981. Ergativity and grammatical relations in Kalaw Lagaw Ya (Saibai dialect). *Australian Journal of Linguistics* 1(1). 1–42.

Conrad, Robert J. & Kepas Wogiga. 1991. *An outline of Bukiyip grammar*. Canberra: Pacific Linguistics.

Corbett, Greville. 1991. *Gender*. Cambridge: Cambridge University Press.

Crass, Joachim. 2014. Personennamen im Libido, einer markierten Nominativsprache in Äthiopien. Paper presented at *Linguistisches Forum*, Universität Regensburg on 14.05.2014.

Dammel, Antje & Corinna Handschuh (eds.). 2019. Grammar of Names. [Special Issue]. *Language Typology and Universals* 72(4).

Daniel, Michael & Edith Moravcsik. 2013. The associative plural. In Matthew S. Dryer & Martin Haspelmath (eds.), *The World Atlas of Language Structures Online*. Leipzig: Max Planck Institute for Evolutionary Anthropology. http://wals.info/chapter/36 (checked 28/10/2020).

D'hulst, Yves, Rolf Thieroff & Trudel Meisenburg. this volume. River names: Definite articles and geographical names.
Dixon, R. M. W. 2010. *Basic linguistic theory. Volume 1: Methodology*. Oxford: Oxford University Press.
Foley, William A. 1991. *The Yimas language of New Guinea*. Stanford, CA: Stanford University Press.
Fortune, Reo F. 1942. *Arapesh*. New York: Augustin.
Handschuh, Corinna. 2014. *A typology of Marked-S languages*. Berlin: Language Science Press.
Handschuh, Corinna. 2017. Nominal category marking on personal names: A typological study of case and definiteness. In Tanja Ackermann & Barbara Schlücker (eds.), *The morphosyntax of proper names*. [Special issue]. *Folia Linguistica* 51(2). 483–504.
Handschuh, Corinna. 2019. The classification of personal names: A crosslinguistic study of sex-specific forms, classifiers and gender marking on personal names. In Antje Dammel & Corinna Handschuh (eds.), *Grammar of names*. [Special issue]. *Language Typology and Universals* 72(4). 539–572.
Harrienhausen, Bettina. 1990. *Hmong Njua: Syntaktische Analyse einer gesprochenen Sprache mithilfe datenverarbeitungstechnischer Mittel und sprachvergleichende Beschreibung des südostasiatischen Sprachraums*. Tübingen: Niemeyer.
Haywood, Graham. 1996. A Maleu grammar outline and text. In Malcome Ross (ed.), *Studies in languages of New Britain and New Ireland*, vol. 1, 145–196. Canberra: Pacific Linguistics.
Helmbrecht, Johannes. this volume. Proper names with and without definite articles: preliminary results.
Helmbrecht, Johannes, Damaris Nübling & Barbara Schlücker (eds.). 2017. *Namengrammatik* (Linguistische Berichte – Sonderheft 23). Hamburg: Buske.
Himmelmann, Nikolaus P. 1997. *Deiktikon, Artikel, Nominalphrase: Zur Emergenz syntaktischer Strukturen*. Tübingen: Niemeyer.
Kieviet, Paulus. 2017. *A grammar of Rapa Nui*. Berlin: Language Science Press.
König, Christa. 2006. Marked nominative in Africa. *Studies in Language* 30(4). 655–732.
Mańczak, Witold. 1995. Morphologie des noms: Règles de flexion, systèmes de flexion. In Ernst Eichler, Gerold Hilty, Heinrich Löffler, Hugo Steger & Ladislav Zgusta (eds.), *Namenforschung. Ein internationales Handbuch zur Onomastik* (Handbücher zur Sprach- und Kommunikationswissenschaft 11), vol. 1, 427–431. Berlin & New York: Walter de Gruyter.
Mauri, Caterina & Andrea Sansò. 2019. Nouns & co. Converging evidence in the analysis of associative plurals. In Antje Dammel & Corinna Handschuh (eds.), *Grammar of names*. [Special issue]. *Language Typology and Universals* 72(4). 603–626.
Moravcsik, Edith. 2003. A semantic analysis of associative plurals. *Studies in Language* 27(3). 469–503.
Mosel, Ulrike & Ruth Spriggs. 2000. Gender in Teop (Bougainville, Papua New Guinea). In Barbara Unterbeck & Matti Rissanen (eds.), *Gender in grammar and cognition*, 321–349. Berlin: Mouton de Gruyter.
Moyse-Faurie, Claire & John Lynch. 2004. Coordination in Oceanic languages and Proto-Oceanic. In Martin Haspelmath (ed.), *Coordinating constructions*, 445–497. Amsterdam & Philadelphia: John Benjamins.
Nübling, Damaris. 2017. The growing distance between proper names and common nouns in German: On the way to onymic schema constancy. In Tanja Ackermann & Barbara

Schlücker (eds.), *The morphosyntax of proper names*. [Special issue]. *Folia Linguistica* 51(2). 341–367.

Nübling, Damaris, Fabian Fahlbusch & Rita Heuser. 2015. *Namen: Eine Einführung in die Onomastik*. 2nd edn. Tübingen: Narr.

Ozanne-Rivierre, Françoise. 1979. *Textes Nemi (Nouvelle-Calédonie). Volume 2: Bas-Coulna et Haut-Coulna*. Paris: Editiones Peeters.

Piper, Nick. 1989. *A sketch grammar of Meryam Mir*. Canberra: Australian National University MA thesis.

Quirk, Randolph, Sidney Greenbaum, Geoffrey Leech & Jan Svartvik. 1972. *A grammar of contemporary English*. London: Longman.

Salaberri, Iker. this volume. D-marking on Basque personal names from a synchronic and diachronic perspective.

Sasse, Hans-Jürgen. 1993. Das Nomen - eine universale Kategorie? *Language Typology and Universals* 46(3). 187–221.

Schachter, Paul & Timothy Shopen. 2007. Parts-of-speech systems. In Timothy Shopen (ed.), *Language typology and syntactic description. Volume I: Clause structure*, 1–60. 2nd edn. Cambridge: Cambridge University Press.

Schlücker, Barbara & Tanja Ackermann. 2017. The morphosyntax of proper names: An overview. In Tanja Ackermann & Barbara Schlücker (eds.), *The morphosyntax of proper names*. [Special issue]. *Folia Linguistica* 51(2). 309–339.

Shee, Naw Hsar. 2008. *A descriptive grammar of Geba Karen*. Chiang Mai: Payap University MA thesis.

Stolz, Thomas, Sander Lestrade & Christel Stolz. 2014. *The crosslinguistics of zero-marking of spacial relations* (Studia Typologica 15). Berlin: Mouton de Gruyter.

Stolz, Thomas and Nataliya Levkovych. this volume. On *Special Onymic Grammar (SOG)*: Definiteness markers in Fijian and selected Austronesian languages.

Stolz, Thomas, Nataliya Levkovych & Aina Urdze. 2017. Die Grammatik der Toponyme als typologisches Forschungsfeld: Eine Pilotstudie. In Johannes Helmbrecht, Damaris Nübling & Barbara Schlücker (eds.), *Namengrammatik* (Linguistische Berichte – Sonderheft 23), 121–146. Buske: Hamburg.

Van Breugel, Seino. 2008. *A grammar of Atong*. Melbourn: La Trobe University PhD thesis.

Van de Velde, Mark L. O. 2006. Multifunctional agreement patterns in Bantu and the possibility of genderless nouns. *Linguistic Typology* 10(2). 183–221.

Van de Velde, Mark L. O. 2008. *A grammar of Eton*. Berlin: Mouton de Gruyter.

Van Langendonck, Willy. 2007. *Theory and typology of proper names*. Berlin: Mouton de Gruyter.

Van Langendonck, Willy & Mark Van de Velde. 2016. Names and grammar. In Carole Hough & Daria Izdebska (eds), *The Oxford handbook of names and naming*, 17–38. Oxford: Oxford University Press.

von Prince, Kilu. 2015. *A grammar of Daakaka*. Berlin: Mouton de Gruyter.

Willett, Thomas L. 1991. *A reference grammar of Southeastern Tepehuan*. Dallas, TX: Summer Institute of Linguistics.

Javier Caro Reina
The definite article with personal names in Romance languages

Abstract: This paper examines the factors that motivate the occurrence of the definite article with personal names in Romance languages from a synchronic and diachronic perspective. The Romance languages selected are Asturian, Catalan, French, Galician, Italian, Occitan, Portuguese, Romanian, Sardinian, Spanish, and Sursilvan. It will be demonstrated that a differentiation of semantic definite contexts in terms of dimensions of knowledge contributes to a better understanding of the use of definite article with famous and ordinary names. The definite article is employed with famous and ordinary names in Balearic Catalan, Galician, Italian, Brazilian Portuguese, European Portuguese, and Sursilvan. The definite article is restricted to famous names in Asturian, French, Sardinian, and Spanish while it is restricted to ordinary names in Central Catalan as well as in non-standard varieties of French, Italian, and Spanish. In Romanian, personal names lack the definite article. In addition, the occurrence of the definite article can be triggered by semantic-pragmatic, lexical, morphosyntactic, phonological, and sociolinguistic factors. In French and Spanish, the absence of the definite article with ordinary names is related to a Standard Average European feature. By contrast, in Romanian, the absence of the definite article with personal names resulted from deflection.

Keywords: definite article, personal name, Romance languages.

1 Introduction

Although the presence of the definite article with common nouns has been extensively studied from a typological perspective (Dryer 2013; Haspelmath 2013), little is known about the cross-linguistic patterns of the definite article with proper names. First typological surveys have been undertaken by Handschuh (2017), Helmbrecht (this volume), and Stolz and Levkovych (this volume). For example, on the basis of 34 genetically and areally diverse languages, Hand-

---

Javier Caro Reina: Romanisches Seminar, Universität zu Köln, Albertus-Magnus-Platz, D-50923 Cologne. E-mail: javier.caroreina@uni-koeln.de.

https://doi.org/10.1515/9783110672626-003

schuh (2017: 498–500) shows that the two most frequent patterns involve the absence of the definite article with personal names and common nouns (15 languages) and the presence of the definite article with personal names and common nouns (11 languages). In this respect, personal names behave like common nouns. However, the author observes that the presence of the definite article is restricted to common nouns in six languages (Basque, Fongbe, Jamsay, Sandawe, Saramaccan, and Zialo). In these languages, personal names differ from common nouns. She further points out that this less common pattern predominantly occurs in European and West African languages.

Among Indo-European languages, the use of the definite article with personal names has mostly been studied in Scandinavian languages (Delsing 1993: 53–55; 2003: 20–25; Sigurðsson 2006; Johannessen and Garbacz 2014) and nonstandard varieties of German (see Werth 2020: 102–143 for a comprehensive overview). With regard to Scandinavian languages, the preproprial article is employed in Icelandic, Norwegian, and Northern Swedish, as illustrated in Map 1 (see Delsing 2003: 21–23 for more details).[1] In Icelandic, first names optionally take the personal pronoun *hann* in the masculine (*hann Jón* 'John') and *hún* in the feminine (*hún Anna* 'Anna'). As pointed out by Delsing (2003: 21), this use is typical of informal speech, but not of written language. In Norwegian, first names require the personal pronoun *han* for the masculine (*han Per* 'Peter') and *ho* for the feminine (*ho Kari* 'Carrie'). However, the varieties spoken in Bergen and Sunnhordland feature the postproprial article *-en* for the masculine (*Peren* 'Peter') and *-a* for the feminine (*Karia* 'Carrie'), which coincide with the suffixed definite article of common nouns. In Bergen, the postproprial article is restricted to male personal names (Håberg 2010: 9). In Northern Swedish, first names require the reduced (clitic) personal pronouns *(n)e* in the masculine (*e Erik*) and *a* in the feminine (*a Anna*). The preproprial article is derived from the third-person singular personal pronoun and behaves differently in Icelandic, Norwegian, and Northern Swedish. First, the preproprial article is optional in Icelandic, but obligatory in Norwegian and Northern Swedish. Second, the preproprial article is inflected in Icelandic, but not in Norwegian and Northern Swedish. This implies that the preproprial article is more grammaticalized in Norwegian and Northern Swedish than in Icelandic. Third, in Norwegian the preproprial

---

1 Following Eriksson (1973: 25), Delsing (1993: 54) talks about preproprial articles to refer to the personal pronouns that occur with personal names and kinship names in argumental position (see Delsing 2003: 20). These personal pronouns have been analysed as determiners by Roehrs (2005: 264–265). More recently, Sigurðsson and Wood (2020: 10) have coined the term preproprials (or preproprial constructions).

article can be employed both with ordinary names (*han Olaf*) and with famous names (*han Elvis Presley*). Importantly, in Icelandic, Norwegian, and Northern Swedish we find distinct definite article forms for personal names and common nouns. Examples from Norwegian and Northern Swedish are given in (1) and (2), respectively.

(1) Norwegian
han Per / ho Kari / mann-en / kvinne-n
'Peter / Kari / the man / the woman'

(2) Northern Swedish
e Erik / a Anna / mann-en / kvinna-n
'Erik / Anna / the man / the woman'

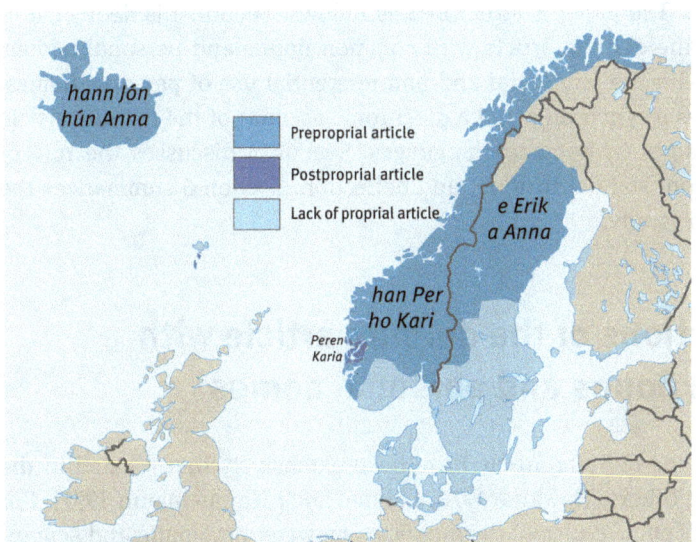

Map 1: The proprial article in Scandinavian languages (adapted from Dahl and Edlund 2019: 71).

Scandinavian languages show that the occurrence of the definite article with personal names has to be examined regarding aspects such as ordinary vs. famous name, gender (male vs. female), personal name class (first vs. family name), obligatoriness (optional vs. compulsory), register (formal vs. informal speech), morphology (case, gender), and distinct article forms for personal names and common nouns.

As for Romance languages, the occurrence of the definite article with personal names has not received much attention. For example, the RAE/ASALE (2009: 840) only mentions that the use of the definite article with first names (*el Manolo, la Juana*) is characteristic of a number of non-standard varieties of Spanish. Similarly, the Bon Usage (Grevisse and Goosse 2016: 834) observes that the definite article is employed with first names especially in rural varieties (*le Gaëtan, la Louise*). Despite research on the definite article with personal names in Catalan (Coromina Pou 2001; Rabella Ribas 2006), Italian (Kubo 2016), Portuguese (Callou and Silva 1997), Romanian (Miron-Fulea et al. 2013), and Spanish (De Mello 1992; García Gallarín 1999), the influence of semantic-pragmatic, lexical, and morphosyntactic factors has not been carefully evaluated in a large sample of Romance languages and language varieties.

The goal of this paper is to examine the factors that condition the occurrence of the definite article with specific personal name classes in selected Romance languages. The paper is structured as follows: Section 2 is dedicated to the functions of the definite article with common nouns and personal names. Section 3 deals with the referential and non-referential use of personal names. Section 4 provides a synchronic and a diachronic account of the definite article with personal names in Romance languages. Section 5 discusses the role of grammaticalization, standardization, and deflection. Section 6 summarizes the main results of the study.

2 The functions of the definite article with common nouns and personal names

The functions of the definite article have been extensively investigated in the literature (among others: Hawkins 1978; Löbner 1985; Himmelmann 1997). For example, Löbner (1985: 298–299) distinguishes between pragmatic and semantic definites. In pragmatic definite contexts, the referent is unambiguously identifiable in the immediate situation or context of utterance. This is the case in deictic, anaphoric, and endophoric uses. By contrast, in semantic definite contexts the referent is unambiguously established regardless of the immediate situation or context of utterance. The difference between pragmatic and semantic definites is useful for accounting for the distribution of the definite article and the demonstrative (Himmelmann 1998: 323), weak and strong forms of the definite article (Schwarz 2009), the grammaticalization of the definite article (de Mulder and Carlier 2011), morphosyntactic phenomena such as contraction of preposition and definite article in German (Löbner 1985: 311–312), and the occurrence of the definite article in the languages of the world (Dryer 2013). For example, in some languages the definite article is restricted to pragmatic definite contexts (as in Garrwa) while in others it is restricted to semantic definite contexts (as in Ma'di) (examples taken from Dryer 2014: e238). In others, there are different article forms for pragmatic and semantic definite contexts, as in German and Germanic dialects. In this respect, Schwarz (2009) talks about strong and weak articles, respectively (see Schwarz 2011 for a typological overview).

The analysis of the functions of the definite article has revolved around common nouns. In comparison, the use of the definite article with proper names has received less attention. For example, as pointed out by Clark and Marshall (1981: 21), Hawkins (1978) does not discuss proper names. In this respect, little is known about the functions of the definite article with proper names (see Sturm 2005 for German). A possible explanation is that previous research has focused on languages such as English, where personal names lack the definite article. With regard to the grammaticalization of the definite article, Greenberg (1978: 64–65) observes that proper names are reluctant to take the definite article since they are inherently definite (see Lyons 1999: 21–22 for discussion).[2] More recently, Löbner (2011: 320–321) put forward a scale of uniqueness which captures the spread of the definite article from pragmatic to semantic definite contexts. The scale of uniqueness is shown in (3), where proper names occupy an intermediate position between inherently unique nouns and personal pronouns. This scale predicts that the presence of the definite article with proper names implies the presence of the definite article with individual and functional concepts and, by extension, that the presence of the definite article with personal pronouns implies the presence of the definite article with proper names (see Stolz and Levkovych this volume for examples from Austronesian languages). Note that proper names are regarded as a homogeneous class. A word of caution, however, is that languages can differ with respect to the use of the definite article with personal names and place names (see Caro Reina and Helmbrecht this volume).

---

2 Interestingly, this pattern persists in Greenberg's Stage III, where the definite article becomes a classifier. This is the case in Bantu languages where proper names are assigned to the so-called 1a class which is characterized by the absence of a classifier (see Van de Velde 2019: 240–241 for details).

(3)   Scale of uniqueness (Ortmann 2014: 314, based on Löbner 2011)
        deictic sortal noun < anaphoric sortal noun < sortal noun with establishing relative clause < definite associative anaphors < individual nouns/functional nouns < proper names < 3rd person personal pronouns < 2nd and 1st person personal pronouns

Importantly, the occurrence of the definite article with personal names in semantic definite contexts provides insights into functions which are absent from common nouns. I will now show that breaking up semantic definite contexts into different dimensions of knowledge can contribute to a better understanding of the use of the definite article with personal names. With regard to semantic definite contexts, scholars have claimed that the universe of discourse or situation can vary (among others: Ebert 1971: 83; Hawkins 1978: 117-118; Leonetti 1999: 798; Szczepaniak 2009: 73). For example, Ebert (1971: 83) points out that the universe of discourse can range from a whole speech community (*a köning* 'the king') to a reduced speech community such as a village (*a sarkklooken* 'the church bells') or a family (*a hünj* 'the dog'). In a similar vein, Hawkins (1978: 115) states that larger situations can be of varying size (country, county, town, and village). In addition, he (1978: 118) differentiates between general and specific knowledge.

Following Ebert (1971) and Hawkins (1978), I will classify semantic definite contexts according to different dimensions and types of knowledge. The dimension of knowledge can refer to the world, a country, a region, a village, and a family (or social group) while the type of knowledge can be general or specific. General knowledge correlates with higher dimensions of knowledge while specific knowledge correlates with lower dimensions of knowledge. I view the types and dimensions of knowledge as a continuum, as can be seen in Figure 1.

	general knowledge				specific knowledge
Dimension of knowledge:	world	country	region	village	family or social group

Figure 2: Semantic definite contexts according to dimension and type of knowledge.

Crucially, the dimensions of knowledge do not have an impact on the occurrence of the definite article with common nouns.[3] By contrast, they can influ-

---

[3] One exception are inherently unique nouns such as *sun, moon*, etc., which are deprived of the definite article in earlier stages of the grammaticalization of the definite article. This is the case in earlier stages of Germanic and Romance languages such as Old High German (Gräf

ence the occurrence of the definite article with personal names. As a consequence, languages where the definite article is sensitive to the dimension of knowledge provide cues about information retrievable from general (or encyclopedic) and specific (or specific-shared) knowledge. Moreover, general knowledge does not necessarily constitute a homogeneous group. As we will see, personal names involving general knowledge can be coded differently depending on whether they refer to the world or a country (see Section 4.2.1).

In what follows, I will provide morphosyntactic evidence supporting the relevance of the dimensions of knowledge.[4] First, the prepropial article is common in Icelandic, Norwegian, and Northern Swedish (see Section 1). However, its occurrence highly depends on the dimension of knowledge since the prepropial article is restricted to ordinary names in Icelandic (*hann Jón* 'John', **hann Elvis Presley*) and Northern Swedish (*han Per* 'Peter', **han Elvis Presley*) while it can appear with ordinary and famous names in Norwegian (*han Olaf, han Elvis Presley*). Second, Cologne Ripuarian exhibits weak and strong articles (see Himmelmann 1997: 54–55).[5] The weak article occurs with ordinary and famous names, as shown in (4). Famous names are illustrated with names of politicians (b), artists (c), sportspeople (d), and historical characters (e).[6] Interestingly, personal names involving specific knowledge (family, social group, village) can take both the weak and strong article while personal names involving general knowledge (country, world) can only take the weak article. In other words, the use of the strong article is sensitive to the dimension of knowledge.

(4) Cologne Ripuarian (Herrwegen, p.c.)
 a. *der/dä Pitter* 'Peter', *et/dat Marie* 'Mary'
 b. *der (Barack) Obama, et Angela Merkel*
 c. *der Dali, der Monet, et Charlotte Link, et Isabel Allende*
 d. *der Messi, der Ronaldo, der Lukas Podolski*
 e. *der Stalin, der Napoleon*

1905: 14–27; Flick 2020: 151–156) and Old French (Gamillscheg 1957: 91–92; Buridant 2000: 109; GGHF 2020: 973, 1549).
4 The examples from the literature point to morphosyntactic differences between ordinary and famous names. However, more research (especially corpus-based studies) is needed in order to fully comprehend the morphosyntactic patterns of ordinary and famous names.
5 The forms that are relevant for the present discussion are the weak articles *der* [də] (masculine) and *et* [ət] (neuter) as well as the strong articles *dä* [dɛː] (masculine) and *dat* [dat] (neuter). Note that in Cologne Ripuarian, feminine names are accompanied by the neuter article.
6 In addition to personal names, deity names occur with the definite article (*der Allah, der Buddha*), with the exception of *God* (*Ø Godd*).

Third, in Classical Latin, the demonstrative *ille* is attested with historical personal names, as in *Cicero ille* 'the famous Cicero' and *Romulus ille* 'the famous Romulus' (examples taken from Marouzeau 1922: 158). Thus, its occurrence is restricted to personal names involving encyclopedic knowledge. Finally, further evidence for the relevance of the domains of knowledge comes from possessive noun phrases in Romanian and Hebrew. In Romanian, possessive constructions involving ordinary names such as *Ion* 'John' are preceded either by the genitive proprial article *lui* or by the preposition *de* 'of'. However, names of famous writers such as *Rebreanu* and *Slavici* can only be preceded by the preposition *de* (Dobrovie-Sorin and Giurgea 2013: 336–337). Similarly, in Hebrew, possessive constructions involving construct states, where possession is overtly coded on the head (possessum), cannot contain ordinary names such as *Yehuda* as possessors in the dependent. One exception are names of famous poets such as *Yehuda Amichai* and *Zelda*. In this respect, Rothstein (2017: 439) observes that the more famous the person denoted by the personal name, the more acceptable the construct state is.

(5) Romanian (Dobrovie-Sorin and Giurgea 2013: 336–337)
    a. *Romanele lui/de Ion*
       'The novels of Ion'
    b. *Romanele *lui/de Rebreanu*
       'The novels of Rebreanu'

(6) Hebrew (Rothstein 2017: 439)
    a. *širey *Yehuda*
       'Poems of Yehuda'
    b. *širey Yehuda Amichai / Zelda*
       'Poems of Yehuda Amichai / Zelda'

In summary, the dimensions of knowledge contribute to a better understanding of the morphosyntactic behaviour of personal names in Icelandic and Norwegian (preproprial article with ordinary names), Cologne Ripuarian (strong article with ordinary names), Latin (demonstrative with famous names), Romanian (*lui* 'of' with ordinary names), and Hebrew (construct states with famous names). In Section 4.1, we will see that the dimensions of knowledge can also have an impact on the occurrence of the definite article with personal names in Romance languages. Importantly, this semantic-pragmatic factor can be accompanied by pragmatic (negative connotation) and morphological (case, gender) factors, as will be shown in Section 4.2.

I will now elaborate on a classification of personal names based on the notion of dimensions of knowledge. I will distinguish between ordinary names and regionally, nationally, and universally famous names, as illustrated in Figure 2. Ordinary names correspond to the dimension of knowledge related to the village, the family, or the social group while regionally, nationally, and universally famous names correspond to the dimension of knowledge related to the region, the country, and the world, respectively. Admittedly, universally famous names constitute a broad category which includes names of current and past celebrities (artists, politicians, philosophers, sportspeople, writers, etc.) as well as fictional, historical, and religious characters. As a consequence, they can morphosyntactically behave in a different way (see Section 4.2.1).

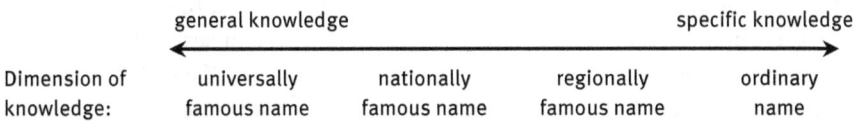

	general knowledge			specific knowledge
Dimension of knowledge:	universally famous name	nationally famous name	regionally famous name	ordinary name

Figure 3: Classification of personal names according to type and dimension of knowledge.

It should be noted that this classification is not viewed as an implicational scale. In this respect, Kokkelmans (2018: 81) set up a hierarchy that captures the occurrence of the definite article with names of deities (*Yahweh*), sacred figures (*Mary*), celebrities or fictional characters (*Elvis Presley*), and personally known people (*Jane*) in Germanic and Scandinavian languages. He assumes an implicational hierarchy according to which the presence of the definite article with famous names implies the presence of the definite article with ordinary names, but not vice versa. However, this unidirectional implication does not hold cross-linguistically. As we will see in Section 4, in some Romance languages the definite article is restricted to famous names while in others it is restricted to ordinary names. This issue will be discussed in more detail in Section 5.

3 Referential vs. non-referential use

The occurrence of the definite article with personal names depends on whether personal names involve refential or non-referential uses.[7] The definite article can appear in the referential use while it is generally excluded from the non-referential use. The examples of definite article with proper names presented so far involve the unmodified referential use. Note that the referential use also includes copular clauses. In this respect, Mikkelsen (2005) provides a compositional semantic analysis of the four clause types previously put forward by Higgins (1979: 204–293): predicational (*Susan is a doctor*), specificational (*The winner is Susan*), equative (*She is Susan*), and identificational (*That is Susan; That woman is Susan*). Mikkelsen (2005: 48–50) demonstrates that the personal name *Susan* is referential, both as the subject in predicational clauses and as the predicate complement in specificational, equative, and identificational clauses.[8] Morphosyntactic evidence supporting this compositional semantic analysis comes from languages where personal names take the definite article.[9] An example from Central Catalan is shown in (7), where the personal name *Maria* 'Mary' requires the definite article in the four different copular clause types.

(7) Central Catalan
    a. Predicational: *La Maria és doctora*
    b. Specificational: *La guanyadora és la Maria*
    c. Equative: *Ella és la Maria*
    d. Identificational: *Aquesta és la Maria; Aquesta dona és la Maria*

---

7 I follow Fernández Leborans (1999: 103) and Raposo and Bacelar (2013: 1013–1017) for a distinction between referential, denominative, and vocative use. Similarly, Delsing (1993) distinguishes between argumental and non-argumental position.
8 Similarly, Raposo and Bacelar (2003: 1014–1015) view personal names in copular sentences in Portuguese as referential arguments. In contrast, Fernández Leborans (1999: 106) adopts an opposing view for Spanish. A word of caution is that the absence of the definite article with personal names in languages such as Spanish does not necessarily imply that personal names are not referential in copular sentences.
9 Such is the case in non-standard German (Werth 2020: 275–279), Greek (Holton et al. 2012: 357, 361), Icelandic (Sigurðsson 2006: 225–226), Norwegian (Delsing 2003: 21), and Northern Swedish (Delsing 1993: 54–55). Interestingly, identificational copular clauses may behave differently. For example, Bellmann (1990: 274) observes that in Dresden, Erfurt, and Leipzig, the definite article is employed when the referent is pointed at in a photograph (*Das ist der Peter* 'This is Peter'), but not when the referent is personally introduced to the addressee (*Das ist ∅ Peter* 'This is Peter').

Referential arguments can be subdivided into unmodified and modified arguments. Modification includes adjective phrases, prepositional phrases, and relative clauses. Some languages avoid modification of proper names (Dixon 2010: 102, 108). In contrast, other languages allow modification of proper names. However, when modified they can morphosyntactically behave like common nouns (see Caro Reina and Helmbrecht this volume). Romance languages generally allow modification of proper names. This is the case in Asturian, Catalan, French, Galician, Italian, Portuguese, Sardinian, Spanish, and Sursilvan.[10] However, we find syntactic restrictions in Central Catalan and Romanian. In Central Catalan, where masculine personal names such as *Joan* 'John' take the onymic article *en*, modification is only possible in postnominal position. This implies that adjectives can occur in postnominal position (*en Joan petit* 'little John'), but not in prenominal position (*en petit Joan 'little John'). Interestingly, the onymic article is preserved with modification involving adjectives, prepositional phrases, and relative clauses, as shown in (8).[11] That is, modified personal names do not behave like common nouns.

(8)   Central Catalan (Lloret, p.c.)
        en Joan / en Joan petit / en Joan amb barba / en Joan que ha respòs.
        'John / little John / the John with the beard / the John that answered.'

In Romanian, personal names such as *Ion* 'John' exhibit the following restrictions: First, adjectives can occur as prenominal modifiers (*deșteptul Ion* 'smart John'), but not as postnominal modifiers (*Ion deșteptul 'smart John'). In this respect, personal names differ from place names and common nouns, which allow both prenominal modification (*medievalul București* 'Medieval Bucharest', *medievalul oraș* 'medieval city') and postnominal modification (*Bucureștiul medieval* 'Medieval Bucharest', *orașul medieval* 'medieval city').[12]

---

10 See Spescha (1989: 211–212, 215), ALLA (2001: 77), Riegel et al. (2018: 317–319), Cunha and Cintra (2017: 238), and Mannu (2017: 158).
11 Note, however, that scholars such as Longobardi (1994: 657), Coromina Pou (2001: 143–144), and Matushansky (2006: 303–304) assert that in Central Catalan, modification of personal names gives rise to the use of the definite article. However, this is not the case in Gironès Catalan (Lloret, p.c.) and Balearic Catalan (Pons-Moll, p.c.). In this respect, additional research is needed to fully examine the impact that modification can have on the morphosyntactic behaviour of personal names in different varieties of Catalan.
12 In Romanian, the position of the definite article is fixed. More specifically, the definite article is attached to the first constituent of the determiner phrase, regardless of whether it is a

Second, postnominal modification involving adjective phrases, prepositional phrases, and relative clauses is only possible with the determiner *cel* (masc.)/*cea* (fem.). In contrast, common nouns such as *băiat* 'boy' can be directly modified, as in the examples:

(9)   Romanian (adapted from Miron-Fulea et al. 2013: 738)
    a.   *Ion cel deștept / Ion cel cu barbă / Ion cel care a răspuns.*
        'smart John / the John with the beard / the John that responded.'
    b.   *băiatul deștept / băiatul cu barbă / băiatul care a răspuns.*
        'the smart boy /the boy with the beard / the boy that responded.'

Non-referential arguments include the denominative use and the vocative. The denominative use results when personal names occur in isolation (lists, anthroponymic dictionaries), in metalinguistic use, and as arguments of predicative verbs such as *call*. In this respect, scholars have coined terms such as "disembodied names" (Gardiner 1954: 8), "nomination" (Jonasson 1994: 69), and "uso denominativo" (Fernández Leborans 1999: 110–111). The metalinguistic use of a personal name is illustrated in (10) with an example from Galician. In varieties of Galician where personal names take the definite article, the presence of the definite article (*o Antonio, o Manuel*) is associated with a referential use while its absence is associated with a non-referential one. Personal names occurring without the definite article (*Antonio, Manuel*) refer to themselves in denominative contexts.

(10)   Galician (Sousa Fernández 1994: 313–314)
    a.   *O Antonio é mais bonito có Manuel.*
        'Antonio is more beautiful than Manuel.'
    b.   *Antonio é mais bonito ca Manuel.*
        '(The name) Antonio is more beautiful than (the name) Manuel.'

I will now illustrate the absence of the definite article in non-referential arguments involving predicative verbs (*call, baptize*, etc.) and the vocative with examples from European and Brazilian Portuguese, where personal names take the definite article in referential use (*o João* 'John', *a Maria* 'Mary'). With regard to denominative arguments, the definite article is not required in European Portuguese (*ele chama-se João* 'his name is John', *ela chama-se Maria* 'her name

---

noun or an adjective (Pană Dindelegan 2013: 241; see Cornilescu and Nicolae 2011: 194–199 for a syntactic analysis).

is Mary') and in Brazilian Portuguese (*ele se chama João* 'his name is John', *ela se chama Maria* 'her name is Mary') (Raposo and Bacelar 2003: 1015).[13] In the vocative, the definite article does not occur with personal names in Brazilian Portuguese (*João!* 'John!', *Maria!* 'Mary!'). By contrast, in European Portuguese the definite article can occur in deferential contexts, both with personal names (*o João!* 'John!', *a Maria!* 'Mary!') and with personal names preceded by titles (*o senhor João!* 'Mr John!', *a senhora Maria!* 'Ms Mary'), as illustrated in (11). Note that this use is also attested in Andean, Caribbean, and Chilean Spanish, albeit in the combination of title and personal name (RAE/ASALE 2009: 1259). In non-deferential contexts, European Portuguese behaves like Brazilian Portuguese (*João!* 'John!', *Maria!* 'Mary!'). While the use of the definite article with common nouns in the vocative is common in Romance languages such as French and Sardinian (Blasco Ferrer 1986: 94; Riegel et al. 2018: 310, 776), the use of the definite article with proper names in the vocative is restricted to European Portuguese.

(11)   European Portuguese (Cintra 1972: 15)
        o António! / o Sr. António! / a Maria! / a Sr.ª Maria!
        'António! / Mr António! / Mary! / Ms Mary!'

4 Survey of Romance languages

In this section, I will provide a synchronic and a diachronic account of the use of the definite article with personal name classes in Romance languages. The selected languages are Asturian, Catalan, French, Galician, Italian, Occitan, Portuguese, Romanian, Sardinian, Spanish, and Sursilvan. The sources include reference grammars and historical grammars (see Appendix 1). In addition, I used the following linguistic atlases: *Atlas Lingüístico Galego* (ALGa), *Atles Lingüístic del Domini Català* (ALDC), and *Atlas linguistique et ethnographique de la Gascogne* (ALG). Note that linguist atlases document the definite article with ordinary names, but not with famous names. The survey is restricted to personal names in unmodified referential use. The focus will be on first names and family names since nicknames generally take the definite article.[14] Personal names

---

13 Alternatively, Brazilian and European Portuguese also employ *o nome dele é João* 'His name is John' (or *o nome dela é Maria* 'her name is Maria').
14 This is apparent from languages and language varieties which do not feature the definite article with first names and family names such as Asturian (ALLA 2001: 77), Galician (Álvarez

employed as common nouns to metonymically designate the work of an artist were also excluded.[15]

4.1 Results

In Asturian, the definite article does not occur with first names (*Xuan, María*) and family names (*Olaya*) (ALLA 2001: 77). However, it occurs with famous feminine first names in derogative use (*la Letizia, la Terelu, la Cherines*) (Prieto Entrialgo, p.c.). In Medieval Asturian, the use of the definite article with personal names is not attested (García Arias 1988: 161–165).

Catalan dialects behave differently with respect to the definite article with personal names (see Table 1). Valencian and Alghero Catalan do not employ the definite article with personal names. North-Western Catalan uses the definite article *el/lo* for consonant-initial masculine names (*el/lo Pere*), *la* for consonant-initial feminine names (*la Maria*) and *l'* for vowel-initial masculine and feminine names (*l'Antoni, l'Antònia*).[16] Roussillon Catalan employs the onymic marker *en* with masculine names (*en Pere*), but the definite article *la* with feminine names (*la Maria*) and *l'* with vowel-initial masculine and feminine names (*l'Antoni, l'Antònia*). Central Catalan differs from Roussillon Catalan in that consonant-initial masculine names take either the definite article *el* (*el Joan*) or the onymic marker *en* (*en Joan*). These forms can alternate (*El/En Jaume ho sap* 'Jaume knows it') or depend on the linguistic area (see below). Balearic Catalan exhibits the onymic marker *en* for consonant-initial masculine names (*en Pere*), *na* for consonant-initial feminine names (*na Maria*), and *n'* for vowel-initial masculine and feminine names (*n'Antoni, n'Antònia*). The onymic markers *en/na* are etymologically derived from the deferential titles DŎMĬNE 'Mr' and DŎMĬNA 'Ms', respectively (see Caro Reina 2014). In the literature, they are traditionally called "personal articles" (*article personal*) or "onymic articles" (*article onomàstic*).

---

and Xove 2002: 380), Spanish (RAE/ASALE 2009: 845), and Valencian (Colomina Castanyer 2002: 548).

15 Metonymy allows for the occurrence of the definite (and indefinite) article in languages where personal names lack the definite article. This is the case in French (*le Picasso*), Romanian (*Picasso-ul*), and Spanish (*el Picasso*) (Miron-Fulea et al. 2013: 743; Pană Dindelegan 2013: 280; RAE/ASALE 2009: 843; Riegel et al. 2018: 319).

16 Note that in western varieties (*ribagorçà*) the definite article is not commonly used with personal names.

Table 1: Use of the onymic marker and definite article with personal names in Catalan dialects.

Variety	Consonant-initial		Vowel-initial	
	Masculine	**Feminine**	**Masculine**	**Feminine**
Valencian	Ø Pere	Ø Maria	Ø Antoni	Ø Antònia
Alghero Catalan	Ø Pietro	Ø Maria	Ø Antoni	Ø Antonia
North-Western Catalan	el/lo Pere	la Maria	l'Antoni	l'Antònia
Roussillon Catalan	en Pere	la Maria	l'Antoni	l'Antònia
Central Catalan	el/en Pere	la Maria	l'Antoni	l'Antònia
Balearic Catalan	en Pere	na Maria	n'Antoni	n'Antònia

Importantly, Balearic Catalan exhibits distinct forms for personal names and common nouns (*en* vs. *es*, *na* vs. *sa*), as illustrated in (12). In this respect, Balearic Catalan resembles Icelandic, Norwegian, and Northern Swedish (see Section 1). In Central Catalan, this only holds for masculine personal names, as can be seen in (13).

(12)   Balearic Catalan
       en Joan / na Maria / es noi / sa noia
       'John / Mary / the boy / the girl'

(13)   Central Catalan
       en Joan / la Maria / el noi / la noia
       'John / Mary / the boy / the girl'

The use of the definite article with first names has been surveyed by the *Atles Lingüístic del Domini Català* (VIII, Maps 1901–1903). Map 2 documents the behaviour of consonant-initial masculine and feminine first names. In Central Catalan, *en* occurs in eastern varieties (*gironès*) while *na* is attested only in two survey sites (36, 47). With regard to gender, the definite article is restricted to feminine first names in survey sites 128–131, 135–136, 145–146 (Ø *Pere, la María*), but to masculine first names in survey site 117 (*lo Pere*, Ø *María*). Interestingly, the occurrence of the definite article with first names can be lexically conditioned. For example, in survey site 141 there is lexically conditioned variation with feminine first names: *Joana* never occurs with the definite article while *Maria* always takes it. Other names do not behave homogeneously.

I will now discuss Central Catalan and Balearic Catalan in more detail. In Central Catalan, the definite article/onymic marker is employed with ordinary first names (*en Joan, la Maria*), hypocoristics (*en Quim, la Fina*), family names (*en Ferrer, la Ferrer*), and first names followed by family names (*en Joan Ferrer, la Maria Ferrer*). Names of public, historical, mythological, and biblical personalities do not take the definite article (*Prat de la Riba, Pompeu Fabra, Aristòtil, Moisès*) (GEIEC 2018: Ch. 8.4, 10.3). Balearic Catalan resembles Central Catalan with respect to the use of the definite article with personal names. However, Balearic Catalan differs from Central Catalan in two respects. First, famous names are accompanied by the onymic marker, as shown in (14). This is the case with names of politicians (a), writers (b), singers (c), and historical characters (d). Second, the onymic marker also occurs with animal names and place names referring to capes, coves, beaches, reefs, etc. (see Caro Reina 2014: 195–197).

(14) Balearic Catalan (Pons-Moll, p.c.)
   a.   *n' Obama, en Torra, en Felipe González*
   b.   *en Josep Pla, en Ruiz Safon, n'Eduardo Mendoza*
   c.   *en Joan Manuel Serrat, na Shakira, en Messi*
   d.   *en Napoleón, n'Aristóteles, en Llull, en Roosevelt*

In French, ordinary names do not exhibit the definite article (Riegel et al. 2018: 315). By contrast, first names and family names are accompanied by the definite article in non-standard varieties. This is the case in Picard (e^s *Lwēi* 'Louis', e^s *Pǫl* 'Paul'), Provence (*lou Jàque, la Mariòun*), and Wallon (*lu Colas* '[Ni]colas', *lu Célèstine* 'Celestine') (Arnaud 1920: 272; Remacle 1952: 127; Flutre 1955: 44).[17] Interestingly, the use of the definite article with ordinary names is attested for the seventeenth-century. Some examples are given in (15). This raises the question of why the definite article no longer occurs with ordinary names in modern French. This issue will be addressed in Section 5. With regard to famous names, the definite article can be found with masculine and feminine first names (*le Dante, la Callas*) and family names (*le Tasse, la Champmeslé*) (Grevisse and Goosse 2016: 834–835). Hübner (1892: 28–30) observes that this practice began in the sixteenth and seventeenth centuries and became widespread in the eighteenth and nineteenth centuries. Today, it is restricted to a few names of poets and artists.

---

17 The examples provided by Flutre (1955) only include masculine first names. In Wallon, the singular form of the definite article is *lu (l')*, both for the masculine and the feminine (Remacle 1952: 100).

(15)   17th-century French (Grevisse and Goosse 2016: 834)
   a.   *Le Corneille est jolie quelquefois.*
        'Corneille is sometimes beautiful.'
   b.   *Mme de Simiane, fille de la Sévigné.*
        'Ms de Simiane, daughter of Sévigné.'

In Galician, the definite article occurs with ordinary masculine and feminine first names (*o Cidre, a Andrea*), but not with famous names, unless they involve a negative connotation (*o Lope de Vega, a Pardo Bazán*) (Álvarez and Xove 2002: 380). This practice is more frequent among older speakers. The use of the definite article with masculine first names has been surveyed by the *Atlas Lingüístico Galego* (II, Maps 215–216) (see Louredo Rodríguez 2015: 177–181). According to the ALGa (II, Map 215), the definite article is more commonly used in the provinces Lugo, Ourense, and Pontevedra than in A Coruña (see Map 3). In addition, it has a lower incidence after prepositions, which constitutes an instance of article-drop in unmodified prepositional phrases.[18] Article-drop is optional and leads to alternation.[19] In Medieval Galician, the definite article is not attested with personal names (Fernández Jiménez 1971: 68–74).

---

18 The ALGa (II, Map 216) maps the responses to the items *A Miguel collérono preso* 'They took Miguel prisoner', *o de Miguel* 'the one of Miguel', and *Vamos á taberna do Xusto* 'We are going to Xusto's bar', which contain the first names *Miguel* and *Xusto* preceded by the prepositions *a* 'DOM' and *de* 'of', showing that the definite article has a more restricted area (especially Lugo and Ourense).

19 For example, personal names as direct objects can be realized with either *a*-marker and absence of definite article (*A boa muller ollaba a Pedriño* 'the good woman was observing Pedriño') or absence of *a*-marker and presence of definite article (*Eu xa non amaba o Queitán* 'I no longer loved Queitán') (López Martínez 1999: 556; Cidrás Escáneo 2006: 157).

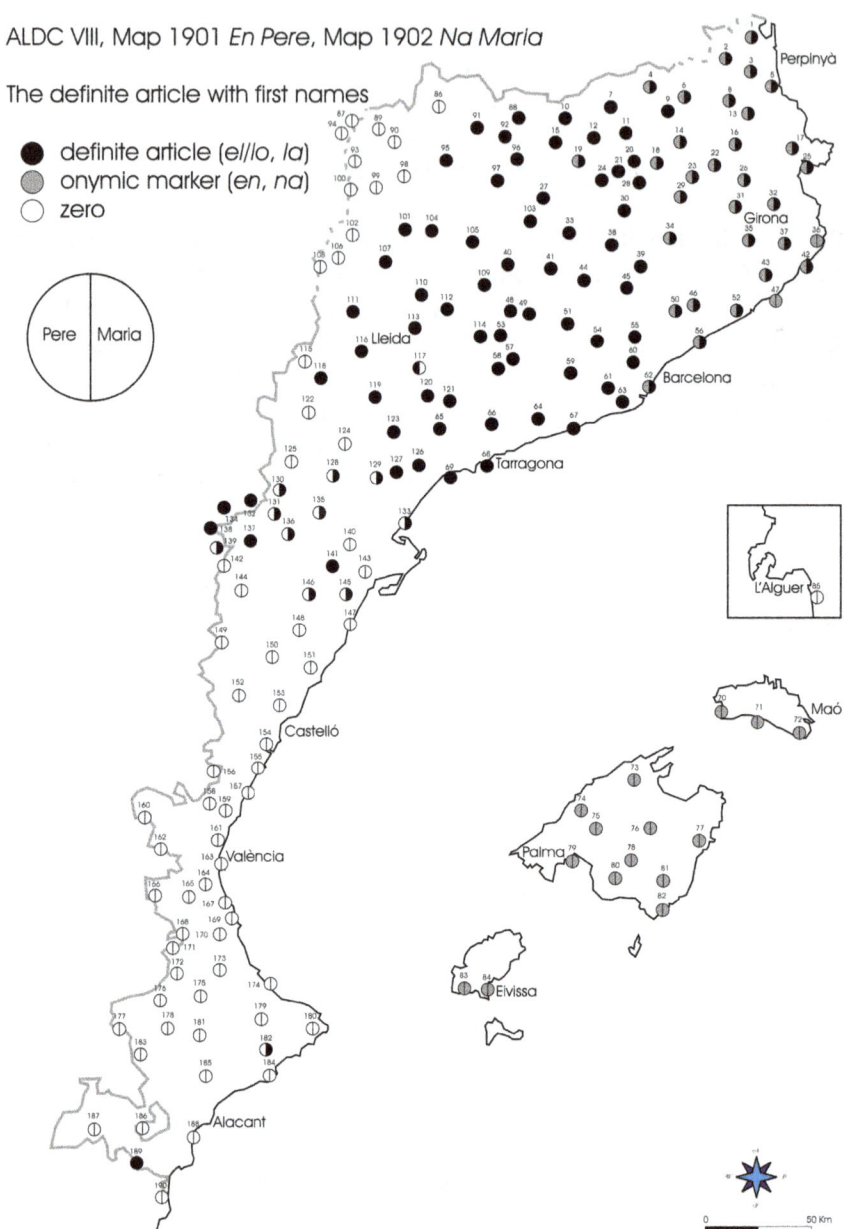

Map 2: The use of the definite article with first names in Catalan.

ALGa II, Map 215
o Xan é o noso veciño 'John is our neighbour'

The definite article with masculine first names

● *o Xan* (with definite article)
○ *Xan* (without definite article)

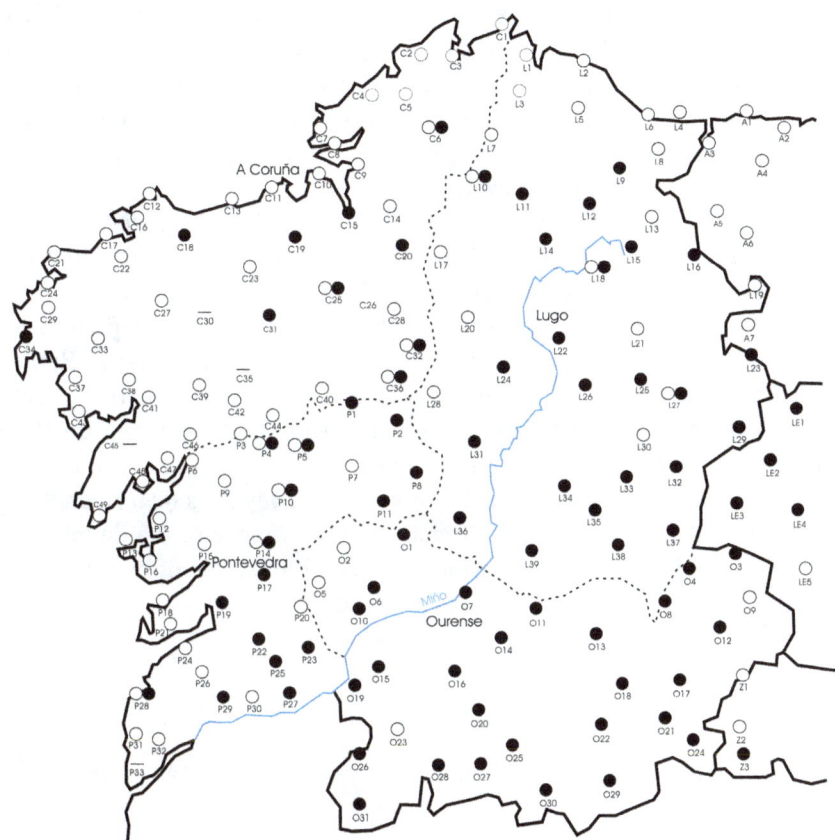

Map 3: The use of the definite article with masculine first names in Galician.

In Italian, the article is optional with ordinary feminine first names (*la Maria*), but obligatory with ordinary feminine family names (*la Corti*) and famous first names and family names (*il Correggio*). By contrast, it is not employed with masculine first names (*Leonardo*), masculine family names (*Sciascia*), the combination of first name and family name (*Leonardo Sciascia, Maria Corti*), deity names (*Dio*, *Dioniso*), and historical names (*Pericle*) (Rohlfs 1969: 29–30; Renzi et al. 2001: 406–407). Kubo (2016) studied the occurrence of the definite article with masculine and feminine first names in Italo-Romance dialects on the basis of the *Atlante Sintattico d'Italia* (ASIt). With regard to gender, the definite article occurs with both masculine and feminine first names in Liguria, Lombardia, Piemonte, Puglia, Toscana, Trentino, and Sardegna while it occurs only with feminine first names in Emilia-Romagna and Veneto. However, the definite article can also be restricted, albeit less frequently, to masculine first names.[20]

In Occitan, the definite article is commonly used with ordinary names. This is the case in Provençal and Gascon, where the definite article is employed with masculine and feminine first names and family names (Ronjat 1937: 123; Joly 1971: 290). Examples from Provençal are *lou Camelot* and *la Gatouno* (Ronjat 1937: 123). The *Atlas linguistique et ethnographique de la Gascogne* (ALG) documents the use of the definite article with masculine and feminine first names (*Jean, Marie*) as well as with masculine family names (*Dupouy*) (VI, Maps 2495–2497).[21] With regard to masculine and feminine first names, the definite article is dominant in Gens, in the north of Hautes-Pyrénées, in the west of Haute Garonne, and in the south of Landes and Lot-et-Garonne (see Map 4). The definite article can occur with both masculine and feminine first names on the one hand, and only with masculine first names on the other. The definite article is confined to feminine first names in only three survey sites (656 Houeillès, 687 Aureilhan, 687E Marseillan). In Occitan, the occurrence of the definite article can be constrained after a preposition as a result of article-drop in unmodified prepositional phrases, which leads to alternations.[22]

---

20 This is the case in some survey sites in Trentino (San Leonardo), Lombardia (Malonno, Semogo), and Liguria (Alassio, Arzeno, Savona).

21 Map 2495 is based on ALG IV, Map 1575 (*C'est Jean qui lira le compliment* 'It's John that will read the speech') and Map 1576 (*Il faudrait bien que Joseph tienne le coup* 'John would have to hold on'). Note that the personal names are in subject position. Map 2497 relies on several responses in the questionnaire.

22 For example, personal names as direct objects can be realized with either *a*-marker and absence of definite article (*Be counechet a Peyou de Mourle?* 'Do you know Peyou de Mourle?') or absence of *a*-marker and presence of definite article (*Be counechet lou Peyou de Mourle?* 'Do you know Peyou de Mourle?') (Joly 1971: 291).

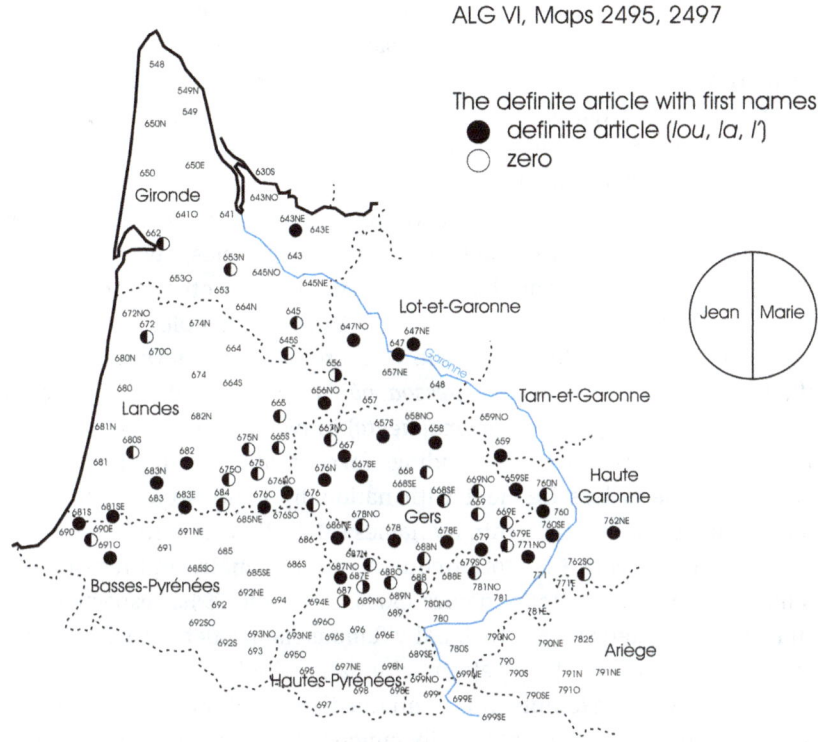

Map 4: The use of the definite article with first names in Gascon.

In European and Brazilian Portuguese, the definite article is generally employed in colloquial register with ordinary names. These include first names (*o João* 'John', *a Maria* 'Mary') and family names (*o Magalhães*) (Thomas 1969: 23–25; Perini 2002: 332–333; Raposo and Bacelar 2003: 1024–1027; Cunha and Cintra 2017: 237–240). In European Portuguese, the definite article is attested with personal names from the fourteenth century onwards although it becomes more frequent in the eighteenth century (Callou and Silva 1997: 14–15). On the basis of the *Projeto da Norma Urbana Linguística Culta*, Silva (1996a, 1996b) studied the use of the definite article in Brazilian Portuguese considering social (age, gender, educational level, etc.) and linguistic factors (personal name class, ordinary vs. famous name). She found that this use is common, albeit to varying degrees, in Recife (17%), Salvador (32%), Rio de Janeiro (43%), Porto Alegre (79%), and São Paulo (87%) (see Lima and Moraes 2019 for northern Brazil). As for famous names, European and Brazilian Portuguese behave similarly with respect to universally famous names, which are characterized by the absence of the definite article (*Camões, Dante, Napoleão*), with the exception of Italian names (*o Tasso*). However, European and Brazilian Portuguese behave differently with respect to nationally and regionally famous names. In European Portuguese, they can take the definite article when they are negatively connotated, as in *o Jorge Sampaio* (Raposo and Bacelar 2003: 1026). By contrast, in Brazilian Portuguese, they can take it without negative connotation. The patterns of the definite article with famous and ordinary names have been investigated in a number of studies on Brazilian varieties (among others: Amaral 2007; Alves de Carvalho 2017; Lima and Moraes 2019). For example, Amaral (2007) examined the use of the definite article with ordinary names (*pessoa do meio social*), regionally famous names (*pessoa pública na região*), and nationally famous names (*pessoa famosa nacionalmente*) in two localities of Minas Gerais (Campanha and Minas Novas). His findings reveal that in Campanha the use of the definite article is less frequent with nationally famous names than with regionally famous names and ordinary names (61%, 82%, and 80%, respectively) while in Minas Novas it is more frequent with nationally and regionally famous names than with ordinary names (50%, 56%, and 29%, respectively), as shown in Table 2. Interestingly, regionally famous names pattern with ordinary names in Campanha while they pattern with nationally famous names in Minas Novas. This is a prime example of the intermediate position that regionally famous names can take between nationally famous names and ordinary names.

Table 2: The use of the definite article with famous and ordinary names in Brazilian Portuguese (adapted from Amaral 2007: 123).

Locality	Nationally famous name	Regionally famous name	Ordinary name
Campanha	61% (43/70)	82% (52/63)	80% (88/109)
Minas Novas	50% (32/63)	56% (13/23)	29% (44/147)

Romanian has an enclitic definite article, which agrees in gender, number, and case with the noun.[23] In the nominative-accusative, personal names lack the

---

23 With regard to case, Romanian distinguishes between nominative-accusative and genitive-dative. The nominative-accusative has the forms *-(u)l*, *-le*, *-a* for the masculine singular and *-a*, *-ua* for the feminine singular. For example, *băiat* 'boy' and *fată* 'girl' are bare nouns while *băiatul* 'boy' and *fata* 'girl' are definites. The genitive-dative has the forms *-(u)lui* for the definite article. In the genitive-dative, the proprial article *lui* is obligatory with masculine names (*lui Ion* 'of John') and feminine names ending in a consonant (*lui Carmen* 'of Carmen'). However, feminine names ending in *-a* take the definite article in formal speech (*Mariei* 'of Maria') or the proprial article *lui* in colloquial (or non-standard) speech (*lui Maria* 'of Maria') (Dobrovie-Sorin et al. 2013: 13–14, 20; Miron-Fulea et al. 2003: 724; Pană Dindelegan 2003: 263–265). The form *lui* has been analysed as a proprial article, which occurs with personal names, kinship names with and without the suffixal possessive (*lui mama* 'of mom', *lui frate-meu* 'of my brother'), animal names (*lui Rex* 'of Rex'), months (*lui martie* 'of March'), letters (*lui a* 'of a'), and numbers (*lui trei* 'of three') (Dobrovie-Sorin et al. 2013: 14; Miron-Fulea et al. 2003: 725). Thus, the definite article only attaches to female personal names ending in *-a* in the genitive-dative. In this respect, personal names differ from place names and common nouns, which require the definite article in the nominative and genitive-dative, as illustrated in (16a)–(16b), and (16c)–(16d), respectively.[24] Moreover, the patterns of the definite article with personal names, place names, and common nouns constitute an instance of morphosyntactic dissociation (see Nübling 2005).

(16) Romanian
    a.   *Ion∅ / Bucureștiul / băiatul*
          'John / Bucharest / the boy'
    b.   *lui Ion / Bucureștiului / băiatului*
          'of/to John / of/to Bucharest / of/to the boy'
    c.   *Carmen∅ / România / fata*
          'Carmen / Romania / the girl'
    d.   *lui Carmen / României / fetei*
          'of/to Carmen / of/to Romania / of/to the girl'

---

masculine singular and *-i* for the feminine singular. The plural forms are not relevant for the present discussion.

24 In the accusative, human definite common nouns display alternation: either with the definite article as in the nominative (*Doctorul examinează băiatul* 'the doctor examines the boy') or without it, but differentially marked and clitic doubled (*Doctorul îl examinează pe băiat* 'the doctor examines the boy') (see Chiriacescu and von Heusinger 2010: 302). Human definite direct objects are differentially marked by means of *pe* 'DOM', giving rise to *pe Ion* 'DOM John', *pe Carmen* 'DOM Carmen', *pe băiat* 'the boy', and *pe fată* 'DOM the girl'. Note that common nouns typically lack the definite article as a result of article-drop in unmodified prepositional phrases. Exceptions are inherently unique nouns such as *împărat* 'emperor', *rege* 'king', etc., which can optionally take the definite article, as in *pe împărat(ul)* 'DOM the emperor', *pe rege(le)* 'DOM the king', etc. (AR 2008: 77).

Table 3 summarizes the inflection of personal names and human common nouns. In modern Romanian, personal names are characterized by onymic schema constancy, according to which the shape of proper names is preserved in order to enable their recognition and processing (see Nübling 2017 for German). First, personal names avoid inflection. More specifically, they are not inflected in the nominative-accusative while they employ the preposition *lui* in the genitive-dative. Second, personal names cannot be modified postnominally by adjective phrases, prepositional phrases, and relative clauses, as shown in (9) above. Third, personal names do not undergo morphophonological alternations typical of common nouns.[25]

Table 3: Definite article with personal names and human common nouns in Romanian.

Case	Personal name		Human common noun	
	Masculine	Feminine	Masculine	Feminine
Nominative	-∅	-∅	-(u)l, -le, -a	-a, -ua
Accusative	-∅	-∅	(-ul), (-le), (-a)	(-a), (-ua)
Genitive-Dative	-∅	-∅, (-i)	-(u)lui	-i

By contrast, in Old Romanian, masculine personal names are attested with the definite article, both in the nominative-accusative (*Radul* 'Radu') and in the genitive-dative (*Radului* 'of Radu') (Pană Dindelegan 2016: 292, 323). As a result of deflection, which began in the sixteenth century, *Radul* and *Radului* were gradually replaced by *Radu* and *lui Radu*, respectively. Note that the ending *-u* of personal names such as *Radu* constitutes a remnant of the definite article *-ul*, which is frequently attested until the end of the nineteenth century (Pană Dindelegan 2013: 290). In addition, in Old Romanian, personal names can be modified postnominally by adjectives without the determiner *cel/cela*.[26]

---

25 Compare *lampa/lămpii* 'the lamp', *seara/serii* 'the evening', and *floarea/florii* 'the flower', which display the stem alternation *a/ă, ea/e, oa/o* in the nominative-accusative and genitive-dative, to the feminine personal names *Sanda/Sandei, Leana/Leanei*, and *Floarea/Floarei*, which preserve the stem vowels *a, ea, oa* throughout the paradigm (Pană Dindelegan 2013: 271).
26 In the sixteenth century, restrictive adjective modification shows alternation between the definite article and the determiner *cel/cela*, as in *Toma necredinciosul* and *Toma cel necredincios* 'the doubting Thomas' (examples taken from Pană Dindelegan 2016: 358).

In Sardinian, ordinary first names do not take the definite article, as illustrated in (17). However, the definite article occurs with names of famous actors, politicians, etc. (Jones 1993; Mannu 2017: 158). In Medieval Sardinian, the use of the definite article with personal names is not attested (Blasco Ferrer 2003: 205–206).

(17) Sardinian (Jones 1993: 59; Mannu 2017: 158)
    a.   Ø *Juanne est diventatu sordatu.*
        'John has become a soldier.'
    b.   Ø *Teresa est diventata bella pitzinna.*
        'Teresa has become a lovely girl.'
    c.   *su Rossellini / su De Sica*
        'Rossellini / De Sica.'

In standard European Spanish, the definite article does not occur with ordinary first names and family names (Fernández Leborans 1999: 103). One exception are famous female family names (especially artists), as in *la Caballé* (RAE/ASALE 2009: 840). In contrast to standard European Spanish, the use of the definite article with ordinary names is common in varieties of European and Latin American Spanish.[27] In Chilean Spanish, we find the neuter article *lo* with

---

[27] In Zamora Castilian, the definite article is used with masculine and feminine first names (*el Simeón*, *la María*) (Baz 1967: 74). In Salamanca Castilian, the definite article is regular with masculine and feminine first names (*el Manolu*, *la Tomasa*) (Llorente Maldonado 1947: 161). In Burgos Castilian, the definite article is generally employed with feminine first names (*la Inés*) and, albeit less frequently, with masculine first names (González Ollé 1964: 35). In Cartagena (Murcia), the definite article occurs with masculine and feminine first names (*el Juan*, *la Antonia*) as well as with masculine and feminine hypocoristics (*el Juanico*, *la Lina*) (García Martínez 1986: 119). In Extremadura, the definite article is used with masculine and feminine first names, both traditional (*el Pedro*, *la María*) and foreign (*el Christian*, *la Jennifer*) (Cummins 1974: 105; Montero Curiel 2006: 52). In Sevilla Andalusian, the definite article is restricted to female first names, especially hypocoristics (*la Conchi* < *Concepción*, *la Encarni* < *Encarnación*, *la Mari* < *María*, *la Sete* < *Setefilla*) (personal knowledge). In Canarian Spanish, the definite article is employed with ordinary (*la María*) and famous names (*el Andrés Segovia*, guitarrist) (Almeida and Díaz Alayón 1988: 111).

In Puerto Rico, the use of the definite article with masculine and feminine first names is negatively connotated (*el José*, *la Tomasa*) (Álvarez Nazario 1992: 181). In Santo Domingo (Dominican Republic), the definite article can occur with feminine first names (*la Juana*) and, though less frequently, with masculine first names (*el Casimiro*) (Henríquez Ureña 1975: 225). In Costa Rica (Valle Central), the presence of the definite article involves a negative connotation with first names (Quesada Pacheco 2002: 95). In Ecuador, we find the definite article with feminine first names in La Sierra (the highlands). By contrast, in la Costa first names are not

family names for denoting someone's estate or farm (*lo Aguirre* 'Aguirre's house') (Lenz 1925: 310). This use is derived from the construction *lo de* typical of Argentinian, Bolivian, Chilean, and Guatemalan Spanish (RAE/ASALE 2009: 1084–1085). The use of the definite article with personal names is attested for the period between the fifteenth and seventeenth centuries (Ortiz and Reynoso 2012; Calderón Campos 2015).[28] This raises the question of why this use diminished in the eighteenth century such that it is absent from modern European Spanish. This issue will be addressed in Section 5.

In Sursilvan, the definite article occurs with ordinary masculine and feminine first names (*il Gieri, la Catrina*), ordinary masculine family names (*il Degonda*), and ordinary and famous feminine family names (*la Casura, la Gartmann*), but not with double first names (*Gion Gieri, Margreta Catrina*), biblical names (*Cristus, Giusep*), and famous masculine family names (*Muoth, Mozart*) (Spescha 1989: 208–210).[29] Examples of ordinary masculine and feminine first names with the definite article are given in (18).

(18)   Sursilvan (Spescha 1989: 208)
    a.   *Il Gieri vul ir vinavon a scola.*
        'Gieri wants to continue going to school.'
    b.   *La Catrina ha entschiet in emprendissadi.*
        'Catrina has started an apprenticeship.'

---

accompanied by the definite article (Toscano Mateus 1953: 153). In San Luis (Argentina), the definite article always occurs with feminine first names (*la María*) and feminine hypocoristics (*la Mecha*), but only sometimes with masculine hypocoristics (*el Juancito*). It has derogatory value with masculine first names (*el Francisco*) and masculine and feminine family names (*el González, la Fernández*). In addition, the use of the definite article with feminine first names is socially conditioned since it is employed when referring to women of a lower social status such as servants (Vidal de Battini 1949: 384–385). For Santiago de Chile, De Mello (1992) shows on the basis of the *Habla Culta* project that the definite article is more common with ordinary and famous feminine names than with ordinary and famous masculine names (first names, family names, and hypocoristics). In Rosario (Argentina), the definite article occurs with masculine and feminine first names in informal speech (*el Pedro, la Juana*). In addition, it has a derogatory or ironic value when it occurs with masculine and feminine family names (*el García, la González*) (Donni de Mirande 1968: 153).

28 Calderón Campos (2015) examines the use of the definite article with personal names on the basis of a diachronic corpus analysis, distinguishing between three cases: definite article with anaphorical function, definite article with famous names, and definite article with ordinary names (see Bello and Cuervo 1905: 227–228 for the anaphorical use in Baroque Spanish). These cases are closely related to text type.

29 In Lumnezia and parts of La Foppa, the definite article is not employed with first names, as in *Paul vul saver nuot da quei* 'Paul does not want to know anything about that'.

4.2 Summary

Table 4 summarizes the factors that trigger the use of the definite article in the surveyed Romance languages. We can distinguish between semantic-pragmatic, lexical, morphosyntactic, phonological, and sociolinguistic factors. In what follows, I will discuss the different factors in more detail. Importantly, these factors can operate independently or in combination with each other. For example, in Asturian the definite article is restricted to famous female first names which are negatively connotated. In Romanian, the definite article only occurs with feminine names ending in a vowel in the genitive-dative in formal speech.

Table 4: Factors conditioning the occurrence of the definite article with personal names in Romance languages.

Factor	Language
Semantic/pragmatic	
Ordinary name	Balearic Catalan, Central Catalan, Galician, Italian, Brazilian Portuguese, European Portuguese, Occitan, Sursilvan, varieties of French/Italian/Spanish
Famous name	Asturian, Balearic Catalan, French, Galician, Italian, Brazilian Portuguese, Sardinian, Spanish, Sursilvan
Negative connotation	Asturian, Galician, European Portuguese, varieties of Spanish
Lexical	
Personal name type	Varieties of Catalan
Personal name class	Asturian, Galician, Italian, Sursilvan, varieties of Spanish
Morphosyntactic	
Case	Galician, Occitan, Romanian
Gender	Asturian, Italian, Occitan, Spanish, Sursilvan, varieties of Catalan/Italian/Spanish
Phonological	
Word-initial segment	Roussillon Catalan
Word-final segment	Romanian
Sociolinguistic	
Non-standard	French, Italian, Spanish
Style	Balearic Catalan, Central Catalan, Brazilian Portuguese, European Portuguese, Romanian

4.2.1 Semantic-pragmatic factors

The classification of personal names based on dimensions of knowledge has proved to be particularly valuable for capturing the patterns of the definite article with personal names (see Table 5). First, the definite article is employed with famous and ordinary names in Balearic Catalan, Galician, Italian, Brazilian Portuguese, European Portuguese, and Sursilvan. Second, the definite article only occurs with a few famous names in Asturian, French, Sardinian, and Spanish. Third, the definite article is confined to ordinary names in Central Catalan as well as in non-standard varieties of French, Italian, and Spanish. Finally, personal names typically lack the definite article in Romanian. The absence of the definite article with ordinary names in French and Spanish and with personal names in general in Romanian will be dealt with in Section 5.

Table 5: Use of the definite article with famous and ordinary names.

Language	Famous name	Ordinary name
Balearic Catalan, Galician, Italian, Brazilian Portuguese, European Portuguese, Sursilvan	+	+
Asturian, French, Sardinian, Spanish	+	–
Central Catalan, varieties of French/Italian/Spanish	–	+
Romanian	–	–

As we have seen in Section 2, the category of famous names is comprised of universally, nationally, and regionally famous names. Morphosyntactic evidence supporting this distinction comes from varieties of Brazilian Portuguese, where universally famous names lack the definite article while nationally and regionally famous names take the definite article, albeit to different degrees. Regionally famous names can pattern with either nationally famous names or ordinary names (see Table 2). These findings support the intermediate position of regionally famous names. Interestingly, we can find differences among universally famous names. For example, the names of current celebrities take the definite article (*o Obama*) while names of past celebrities do not (∅ *Churchill*). This suggests that a more fine-grained distinction is needed for universally famous names.

Crucially, in languages where the definite article is restricted to famous or ordinary names, the presence (or the absence) of the definite article helps to instruct the addressee to retrieve a certain piece of given information from their general (or encyclopedic) or specific (or specific-shared) knowledge. For example, in Central Catalan, the presence of the definite article is related to ordinary names (*la Maria*) while the absence of the definite article is related to famous names (∅ *Maria*, as the biblical name). Another example comes from Brazilian Portuguese, where the absence of the definite article gives hints about universally famous names (∅ *Churchill*), as opposed to nationally famous names (*o Lula*) and ordinary names (*o Tiago*), which are accompanied by the definite article. This leads to the conclusion that personal names involving general knowledge can be coded differently depending on whether they refer to the world or a country.

A further pragmatic factor involves negative connotation. The use of the definite article with personal names is negatively connotated in Asturian, Galician, European Portuguese as well as in Puerto Rican and Costa Rican Spanish (see Footnote 27). In Asturian, Galician, and European Portuguese, the negative connotation is associated with famous names while in Puerto Rican and Costa Rican Spanish it is associated with ordinary names.

4.2.2 Lexical factors

With regard to lexical factors, we can distinguish between personal name type and personal name class. Personal name type means that the definite article is employed with some personal names, but not with others. That is, the occurrence of the definite article is lexically conditioned. Such examples were only detected in some survey sites of the ALDC and are restricted to feminine first names (∅ *Joana* vs. *la Maria*). Personal name class has an influence on the occurrence of the definite article in Asturian, Galician, Italian, Sursilvan, and some varieties of Spanish.

4.2.3 Morphosyntactic factors

Morphosyntactic factors include case and gender. With regard to case, we can observe article-drop in unmodified prepositional phrases in Galician and Occitan. This is the case with lexical and functional prepositions. An example of a functional preposition acting as a case marker is *a*, which is employed for differential object marking involving personal names. In Galician and Occitan, there is alternation with personal names as direct objects: *a*-marker and absence of definite article on the one hand, and absence of *a*-marker and presence of definite article on the other (see Footnotes 19 and 22 for examples). Crucially, article-drop in unmodified prepositional phrases is sensitive to proper name class in Romance languages. We find article-drop with personal names in Galician and Occitan, but with place names in French, Italian, Portuguese, Romanian, Sardinian, and Sursilvan (see Caro Reina 2020: 36–40). In addition, in Romanian, the definite article only occurs in the genitive-dative with feminine names ending in a vowel (*Mariei* 'of/to Mary').

The majority of the surveyed Romance languages and language varieties are not sensitive to gender. However, gender influences the occurrence of the definite article in Asturian, Italian, Occitan, Spanish, and Sursilvan as well as in some varieties of Catalan, Italian, and Spanish.

4.2.4 Phonological factors

Phonological factors can have an impact on the occurrence of the definite article in Romanian and the onymic marker in Roussillon Catalan. In Romanian, the word-final segment determines the presence of the suffixed definite article in the case of feminine names ending in a vowel (*Mariei* 'of/to Mary') or the proprial article in the case of feminine names ending in a consonant (*lui Carmen* 'of/to Carmen'). In Roussillon Catalan, the word-initial segment determines the presence of the onymic marker in the case of consonant-initial masculine names (*en Pere*) or the presence of the definite article in the case of vowel-initial masculine names (*l'Antoni*). This is not the case in Balearic Catalan, which exhibits the onymic article regardless of the word-initial segment (*n'Antoni, en Pere*).

4.2.5 Sociolinguistic factors

There are two sociolinguistic factors that condition the occurrence of the definite article with personal names. The first involves the difference between standard and non-standard varieties. In standard French and Spanish, ordinary names do not take the definite article. By contrast, in varieties of French and Spanish, the definite article is commonly used with ordinary names (see Footnote 27 for examples from European and Latin American Spanish). In addition, in standard Italian, the definite article is optional with ordinary feminine first names while in a few non-standard varieties it is employed with both masculine and feminine ordinary names. The second is register. The use of the definite article with personal names is typical of colloquial register (or spoken speech) in Balearic Catalan, Central Catalan, Brazilian Portuguese, European Portuguese, and Romanian.

Sociolinguistic studies devoted to the factors that influence the occurrence of the definite article with personal names are scarce in the literature. We can mention the work of Alves de Carvalho (2017) and Lima and Moraes (2019) on varieties of Brazilian Portuguese, which investigate social (age, gender, educational level, etc.) and linguistic factors (personal name class, ordinary vs. famous name).

5 Discussion

Some particular questions regarding the use of the definite article with personal names are: Can we assume a grammaticalization process for the occurrence of the definite article with personal names? How can we explain the absence of the definite article with ordinary names in French, Romanian, and Spanish?

The occurrence of the definite article with proper names has been related to later stages of the grammaticalization of the definite article (Greenberg 1978; Lyons 1999: 337; see Szczepaniak and Flick 2020 for discussion). In the different grammaticalization pathways of the definite article, proper names have been viewed as a homogeneous group. However, with regard to personal names, we could observe that the use of the definite article is primarily motivated by semantic-pragmatic factors involving the difference between famous and ordinary names which is based on different dimensions of knowledge. Thus, we could assume a grammaticalization process which begins with ordinary names and then expands to famous names. Alternatively, the grammaticalization process begins with famous names and expands to ordinary names. Once this semantic-pragmatic factor is activated, it can be accompanied by additional factors such as negative connotation, personal name class, case, gender, etc. For example, in Asturian, the definite article only occurs with famous female first names which are negatively connoted. As shown in Section 2, Kokkelmans (2018: 81) set up an implicational hierarchy according to which the presence of the definite article with famous names implies the presence of the definite article with ordinary names, but not vice versa. However, this unidirectional implication does not hold for Romance languages. As we have seen in Section 4.1, the definite article is restricted to famous names in Asturian, French, Sardinian, and Spanish while it is restricted to ordinary names in Central Catalan as well as in some varieties of French, Italian, and Spanish (see Table 5). These findings clash with the unidirectionality typical of grammaticalization. In addition, with regard to the use of the definite article with personal names, König (2018: 177–178) points out: "The development of totally redundant uses of definite articles cannot be analyzed as being part of a wide-spread chain of grammaticalization but must be a lateral development".

Here, I argue that the occurrence of the definite article with personal names can be explained in a more satisfactory way in terms of pragmaticalization rather than grammaticalization. The term "pragmaticalization" was coined by Dostie (2004) in order to analyse the emergence of discourse markers in contemporary Québec French which cannot be properly accounted for in terms of grammaticalization. The notion of pragmaticalization, which allows for more than one chain and involves a wide range of pragmaticalized structures, can be adopted for the use of the definite article with personal names in Romance languages. In other words, the spread of the definite article with personal names obeys a pragmatic function which is not subject to a unidirectional pathway and follows different side roads and a complex ramification network. In this respect, personal names considerably differ from place names. The spread of the definite article with place names takes place along a pathway based on semantic features expanding from less prototypical categories (names of rivers and mountains) to more prototypical ones (names of cities and countries) (see Caro Reina 2020). In this respect, personal names and place names should not be viewed as a homogeneous group.

Let us move on to the absence of the definite article with ordinary names in French, Romanian, and Spanish. As shown in Section 4.1, the definite article is absent from ordinary names in standard French and Spanish. Interestingly, in these languages the definite article is found in earlier historical stages and in non-standard varieties. In French, the definite article is attested with personal names in the seventeenth-century. However, the *Académie Française* (1704) stigmatizes the occurrence of the definite article with personal names ("C'èst tres-mal parler, et contre le genie de nostre langue, qui ne souffre point d'articles aux noms propres" [it is very bad speech, and against the genius of our language, which does not suffer at all from the article with proper names]). Similarly, in Spanish, the definite article is attested with personal names between the fifteenth and seventeenth centuries. However, the first grammar of the *Real Academia Española* (1771: 52) published in the eighteenth century bans the use of the definite article with personal names ("no debe decirse: el Pedro, la Maria" [you must not say: the Peter, the Mary]). Thus, we could assume that the absence of the definite article with ordinary names in modern French and Spanish resulted from language prescription and standardisation. However, more research would be needed to clear up this matter. In turn, the definite article was preserved in non-standard varieties. In this respect, we can view the presence of the definite article with ordinary names in non-standard varieties of French and Spanish as a Non-Standard Average European feature (see Seiler 2019 for German).

Modern Romanian is characterized by the lack of the definite article with personal names. In contrast, Old Romanian exhibits the definite article with masculine personal names, both in the nominative-accusative (*Radul* 'Radu') and in the genitive-dative (*Radului* 'of Radu') (Pană Dindelegan 2016: 292, 323). The loss of the definite article resulted from deflection of personal names, which began in the sixteenth century. In this respect, modern Romanian resembles eighteenth-century German, where deflection first applied to personal names before expanding to place names (see Nübling 2012: 234–238; 2017: 344–349).

In summary, the absence of the definite article with ordinary names in French, Romanian, and Spanish has two explanations: (1) (tentatively) the role of language prescription and standardization in the case of French and Spanish; and (2) deflection in the case of Romanian.

6 Conclusions

This paper has provided a synchronic and a diachronic account of the use of the definite article with personal name classes in Romance languages. The findings can be summarized as follows: First, the use of the definite article with personal names is primarily motivated by a semantic-pragmatic factor based on different dimensions of knowledge involving the difference between famous and ordinary names. The definite article is restricted to famous names in Asturian, French, Sardinian, and Spanish while it is restricted to ordinary names in Central Catalan as well as in non-standard varieties of French, Italian, and Spanish. In addition, the definite article is employed with famous and ordinary names in Balearic Catalan, Galician, Italian, Brazilian Portuguese, European Portuguese, and Sursilvan while it is almost absent from Romanian. Second, this semantic-pragmatic factor can be accompanied by additional ones. These include pragmatic (negative connotation), lexical (personal name type and class), morpho-syntactic (case, gender), phonological (word-initial and word-final segment), and sociolinguistic (non-standard, register) factors. Third, the absence of the definite article with ordinary names in French, Romanian, and Spanish has different origins: role of language prescription and standardization (French and

Spanish) and deflection (Romanian). Finally, Romance languages provide insight into the complex spread of the definite article with personal names.

Acknowledgements

I would like to thank Marco García García and Klaus von Heusinger for insightful comments on a previous version of this paper. Special thanks go to Tiago Duarte, Alice Herrwegen, Clara Elena Prieto, Clàudia Pons-Moll, Maria-Rosa Lloret, and Halldór Armann Sigurðsson for discussion on data from Brazilian Portuguese, Cologne Ripuarian, Asturian, Balearic Catalan, Central Catalan, and Icelandic, respectively. My thanks also go to the two anonymous reviewers.

References

Académie Françoise. 1704. *Observations de l'Académie françoise sur les Remarques de M. de Vaugelas*. Paris: Coignard.
ALDC = Joan Veny & Lídia Pons Griera. 2001–2020. *Atles lingüístic del domini català*. Barcelona: Institut d'Estudis Catalans. http://aldc.espais.iec.cat (checked 01.03.2022).
ALG = Séguy, Jean. 1954–1973. *Atlas linguistique et ethnographique de la Gascogne*. Paris: Centre National de la Recherche Scientifique.
ALGa = Francisco Fernández Rei et al. 1990–. *Atlas lingüístico galego*. A Coruña: Fundación Pedro Barrié de la Maza.
ALLA = Academia de la Llingua Asturiana (ed.). 2001. *Gramática de la llingua asturiana*. 3rd edn. Uviéu: Academia de la Llingua Asturiana.
Almeida, Manuel & Carmen Díaz Alayón. 1988. *El español de Canarias*. Santa Cruz de Tenerife: Romero.
Álvarez Blanco, Maria Rosario & Xosé Xove. 2002. *Gramática da lingua galega*. Vigo: Galaxia.
Álvarez Nazario, Manuel. 1992. *El habla campesina del país: Orígenes y desarrollo del español en Puerto Rico*. Rio Piedras: Universidad de Puerto Rico.
Alves de Carvalho, Ana Paula Mendes. 2017. O comportamento linguístico dos jovens de Barra Longa/MG em relação ao uso do artigo definido diante de antropônimos. *Caletroscópio* 5(8). 69–90.
Amaral, Eduardo Tadeu Roque. 2007. A importância do fator intimidade na variação ausência/presença de artigo definido diante de antropônimos. *Veredas – Revista de Estudos Linguísticos* 11(1). 116–127.
AR = Academia Română (ed.). 2008. *Gramatica limbii române. Volum 1: Cuvântul*. București: Editura Academiei Române.
Arnaud, François & Gabriel Morin. 1920. *Le langage de la Vallée de Barcelonnette*. Paris: Champion.
Ayres-Bennett, Wendy (ed.). 2018. *Claude Favre de Vaugelas. Remarques sur la langue françoise* (Descriptions et théories de la langue française 2). Paris: Classiques Garnier.

Baz, José María. 1967. *El habla de la tierra de Aliste*. Madrid: CSIC.
Becker, Laura. 2021. *Articles in the world's languages* (Linguistische Arbeiten 577). Berlin & Boston: de Gruyter.
Bellmann, Günter. 1990. *Pronomen und Korrektur. Zur Pragmalinguistik der persönlichen Referenzformen*. Tübingen: Niemeyer.
Bello, Andrés & Rufino José Cuervo. 1905. *Gramática de la lengua castellana destinada al uso de los americanos*. 9th edn. Paris: Roger y Chervoviz.
Blasco Ferrer, Eduardo. 1986. *La lingua sarda contemporanea: Grammatica del logudorese e del campidanese. Norma e varietà dell'uso. Sintesi storica*. Cagliari: Edizioni della Torre.
Blasco Ferrer, Eduardo. 2003. *Crestomazia sarda dei primi secoli. Volume 1: Testi, grammatica storica, glossario*. Nuoro: Centro Max Leopold Wagner.
Buridant, Claude. 2000. *Grammaire nouvelle de l'ancien français*. Paris: SEDES.
Calderón Campos, Miguel. 2015. El antropónimo precedido de artículo en la historia del español. *Hispania* 98(1). 79–93.
Callou, Dinah & Giselle M.O. e Silva. 1997. O uso do artigo definido em contextos específicos. In Dermeval da Hora (ed.), *Diversidade lingüística no Brasil*, 11–27. João Pessoa: Idéia.
Caro Reina, Javier. 2014. The grammaticalization of the terms of address *en* and *na* as onymic markers in Catalan. In Friedhelm Debus, Rita Heuser & Damaris Nübling (eds.), *Linguistik der Familiennamen* (Germanistische Linguistik 225–227), 175–204. Hildesheim: Olms.
Caro Reina, Javier. 2020. The definite article with place names in Romance languages. In Nataliya Levkovych & Julia Nintemann (eds.), *Aspects of the grammar of names: Empirical case studies and theoretical topics* (LINCOM Studies in Language Typology 33), 25–51. München: LINCOM.
Caro Reina, Javier & Johannes Helmbrecht. this volume. Morphosyntactic contrasts between proper names and common nouns: an introduction.
Chiriacescu, Sofiana & Klaus von Heusinger. 2010. Discourse Prominence and *Pe*-marking in Romanian. *International Review of Pragmatics* 2(2): 298–332.
Cidrás Escáneo, Francisco Antonio. 2006. Sobre o uso da preposicion "a" con OD en galego. *Verba: Anuario galego de filoloxia* 33, 147–174.
Cintra, Lindley. 1972. *Sobre «formas de tratamento» na língua portuguesa*. Lisboa: Livros Horizonte.
Clark, Herbert H. & Catherine K. Marshall. 1981. Definite reference and mutual knowledge. In Aravind K. Joshi, Bonnie L. Webber & Ivan A. Sag (eds.), *Elements of discourse understanding*, 10–63. Cambridge: Cambridge University Press.
Colomina Castanyer, Jordi. 2002. Paradigmes flectius de les altres classes nominals. In Joan Solà, Maria-Rosa Lloret, Joan Mascaró & Manuel Pérez Saldanya (eds.), *Gramàtica del Català Contemporani. Volum 1: Introducció. Fonètica i fonologia. Morfologia*, 535–582. Barcelona: Empúries.
Coromina Pou, Eusebi. 2001. *L'article personal en català. Marca d'oralitat en l'escriptura*. Barcelona: UAB PhD Thesis.
Cornilescu, Alexandra & Alexandru Nicolae. 2011. On the syntax of Romanian definite phrases. Changes in the patterns of definiteness checking. In Petra Sleeman & Harry Perridon (eds.), *The noun phrase in Romance and Germanic. Structure, variation and change*, 193–222. Amsterdam & Philadelphia: John Benjamins.
Cummins, John G. 1974. *El habla de Coria y sus cercanias*. London: Tamesis.
Cunha, Celso & Lindley Cintra. 2017. *Nova gramática do português contemporâneo*. 7th edn. Rio de Janeiro: Lexikon.

Dahl, Östen & Edlund, Lars-Erik (eds.). 2019. *Språken i Sverige. Sveriges Nationalatlas*. 2nd edn. Stockholm: Norstedts Förlagsgrupp.
De Mello, George. 1992. El artículo definido con nombre propio de persona en el español hablado culto contemporáneo. *Studia Neophilologica* 64(2). 221–234.
Delsing, Lars-Olof. 1993. *The internal structure of noun phrases in the Scandinavian languages: A comparative study*. Lund: Lund University PhD thesis.
Delsing, Lars-Olof. 2003. Syntaktisk variation i skandinaviska nominalfraser. In Lars-Olof Delsing, Anders Holmberg & Øystein Aleksander Vangsnes (eds.), *Dialektsyntaktiska studier av den nordiska nominalfrasen*, 11–65. Oslo: Novus.
Dixon, R. M. W. 2010. *Basic linguistic theory. Volume 1: Methodology*. Oxford: Oxford University Press.
Dobrovie-Sorin, Carmen, Ion Giurgea & Donka Farkas. 2013. Introduction: Nominal features and nominal projections. In Carmen Dobrovie-Sorin & Ion Giurgea (eds.), *A reference grammar of Romanian. Volume 1: The noun phrase*, 1–47. Amsterdam & Philadelphia: John Benjamins.
Donni de Mirande, Nélida E. 1968. *El español hablado en Rosario*. Rosario: Universidad Nacional del Litoral.
Dostie, Gaétane. 2004. *Pragmaticalisation et marqueurs discursifs. Analyse sémantique et traitement lexicographique*. Bruxelles: De Boeck-Duculot.
Dryer, Matthew S. 2013. Definite articles. In Matthew S. Dryer & Martin Haspelmath (eds.), *The world atlas of language structures online*. Leipzig: Max Planck Institute for Evolutionary Anthropology. http://wals.info/chapter/37 (checked 01.03.2021).
Dryer, Matthew S. 2014. Competing methods for uncovering linguistic diversity: The case of definite and indefinite articles (Commentary on Davis, Gillon, and Matthewson). *Language* 90(4). e232–e249.
Ebert, Karen H. 1971. *Referenz, Sprechsituation und die bestimmten Artikel in einem nordfriesischen Dialekt (Fering)*. Bräist/Bredstedt: Nordfriisk Instituut.
Eriksson, Ulrik. 1973. *Åselesvenska 2. Målutjämning: modeller för analys och syntes med tillämpningar* (Lundastudier i nordisk språkvetenskap A 24). Lund: Institutionen för nordilka språk.
Fernández Jiménez, María José. 1971. *Estudio histórico del artículo gallego*. Santiago de Compostela: Universidade de Santiago de Compostela PhD thesis.
Fernández Leborans, María Jesús. 1999. El nombre propio. In Ignacio Bosque & Violeta Demonte (eds.), *Gramática descriptiva de la lengua española*, 77–128. Madrid: Espasa Calpe.
Flick, Johanna. 2020. *Die Entwicklung des Definitartikels im Althochdeutschen: Eine kognitivlinguistische Korpusuntersuchung* (Empirically Oriented Theoretical Morphology and Syntax 6). Berlin: Language Science Press.
Flutre, Louis-Ferdinand. 1955. *Le parler picard de Mesnil-Martinsart (Somme). Phonétique. Morphologie. Syntaxe. Vocabulaire*. Genève & Lille: Droz.
Gamillscheg, Ernst. 1957. *Historische Französische Syntax*. Tübingen: Niemeyer.
García Arias, José Lluis. 1988. *Contribución a la gramática histórica de la lengua asturiana y la caracterización etimológica de su léxico*. Uviéu: Universidá d'Uviéu.
García Gallarín, Consuelo. 1999. *El nombre propio. Estudios de historia lingüística española*. Madrid: PatRom.
García Martínez, Ginés. 1986. *El habla de Cartagena*. Murcia: Universidad de Murcia
Gardiner, Alan Henderson. 1954. *The theory of proper names. A controversial essay*. Oxford: Oxford University Press.

GEIEC = Institut d'Estudis Catalans. 2018. *Gramàtica essencial de la llengua catalana.* https://geiec.iec.cat (checked 01.03.2021).
GGHF = Marchello-Nizia, Christiane et al. 2020. *Grande grammaire historique du français.* Berlin & Boston: De Gruyter.
González Ollé, Fernando. 1964. *El habla de la bureba: Introducción al castellano actual de Burgos.* Madrid: Gómez.
Gräf, Heinrich. 1905. *Die Entwicklung des deutschen Artikels vom Althochdeutschen zum Mittelhochdeutschen.* Giessen: Heppeler & Meyer.
Grevisse, Maurice & André Goosse. 2016. *Le bon usage: Langue française.* 16th edn. Bruxelles: De Boeck Supérieur.
Greenberg, Joseph H. 1978. How does a language acquire gender markers? In Joseph H. Greenberg, Charles A. Ferguson & Edith A. Moravcsik (eds.), *Universals of human language. Volume 3: Word structure,* 47–82. Stanford, CA: Stanford University Press.
Håberg, Live. 2010. *Den preproprielle artikkelen i norsk. Ei undersøking av namneartiklar i Kvæfjord, Gausdal og Voss.* Oslo: Universitetet i Oslo MA thesis.
Handschuh, Corinna. 2017. Nominal category marking on personal names: A typological study of case and definiteness. In Tanja Ackermann & Barbara Schlücker (eds.), *The morphosyntax of proper names.* [Special issue]. *Folia Linguistica* 51(2). 483–504.
Haspelmath, Martin. 2013. Definite articles. In Susanne Maria Michaelis, Philippe Maurer, Martin Haspelmath & Magnus Huber (eds.), *The atlas of pidgin and creole language structures.* Oxford: Oxford University Press. https://apics-online.info/parameters/28.chapter.html (checked 01.03.2021).
Hawkins, John A. 1978. *Definiteness and indefiniteness: A study in reference and grammaticality prediction.* London: Croom Helm.
Henríquez Ureña, Pedro. 1975. *El español en Santo Domingo.* 2nd edn. Santo Domingo: Taller.
Helmbrecht, Johannes. this volume. Proper names with and without definite articles: preliminary results.
Higgins, Roger Francis. 1979. *The pseudo-cleft construction in English.* New York: Garland.
Himmelmann, Nikolaus P. 1997. *Deiktikon, Artikel, Nominalphrase: Zur Emergenz syntaktischer Struktur* (Linguistische Arbeiten 362). Tübingen: Niemeyer.
Himmelmann, Nikolaus P. 1998. Regularity in irregularity: Article use in adpositional phrases. *Linguistic Typology* 2(3). 315–353.
Holton, David, Peter Mackridge & Irene Philippaki-Warburton. 2012. *Greek. A comprehensive grammar.* Revised by Vassilios Spyropoulos. 2nd edn. London: Routledge.
Hübner, Hans. 1892. *Syntaktische Studien über den bestimmten Artikel bei Eigennamen im Alt- und Neufranzösischen.* Kiel: Kiel University PhD thesis.
Johannessen, Janne Bondi & Piotr Garbacz. 2014. Proprial articles. *Nordic Atlas of Language Structures (NALS) Journal* 1(1). 10–17.
Joly, André. 1971. Le complément verbal et le morphème *a* en béarnais. Observations sur le genre et la fonction dans les langues romanes. *Zeitschrift für romanische Philologie* 87(3–4). 286–305.
Jonasson, Kerstin. 1994. *Le nom propre. Constructions et interprétations.* Louvain-la-Neuve: Duculot.
Jones, Michael Allan. 1993. *Sardinian syntax.* London: Routledge.
Kokkelmans, Joachim. 2018. Elvis Presley, God and Jane: the Germanic proprial article in a comparative perspective. *Working Papers in Scandinavian Syntax* 100. 64–98.

König, Ekkehard. 2018. Definite articles and their uses. In Daniël Olmen, Tanja Mortelmans & Frank Brisard (eds.), *Aspects of linguistic variation*, 165–184. Berlin & Boston: De Gruyter.

Kubo, Hiroshi. 2016. Presenza o assenza dell'articolo definito davanti al nome proprio di persona in italiano e nei dialetti italiani. *Quaderni di lavoro ASIt* 19. 47–64.

Lenz, Rodolfo. 1925. *La oración y sus partes. Estudios de gramática general y castellana*. Madrid: Revista de Archivos.

Leonetti, Manuel. 1999. El artículo. In Ignacio Bosque & Violeta Demonte (eds.), *Gramática descriptiva de la lengua española*, 787–890. Madrid: Espasa Calpe.

Lima, Alcides Fernandes de & Ronaldo Nogueira de Moraes. 2019. Uso do artigo definido diante de nome próprio nas capitais do norte do Brasil. *Moraes* 54. 69–93.

Llorente Maldonado de Guevara, Antonio. 1947. *Estudio sobre el habla de la Ribera (Comarca salmantina ribereña del Duero)*. Salamanca: Colegio Trilingüe de la Universidad.

Löbner, Sebastian. 1985. Definites. *Journal of Semantics* 4(4). 279–326.

Löbner, Sebastian. 2011. Concept types and determination. *Journal of Semantics* 28(3). 279–333.

Longobardi, Giuseppe. 1994. Reference and proper names: A theory of n-movement in syntax and logical form. *Linguistic Inquiry* 25(4). 609–665.

López Martínez, María Sol (1999): O emprego de *a+CD* na lingua galega falada. In Rosario Alvarez & Dolores Vilavedra (eds.), *Cinguidos por unha arela común: homenaxe ó profesor Xesús Alonso Montero*, vol. 1, 551–563. Santiago de Compostela: USC.

Louredo Rodríguez, Eduardo. 2015. On the use of the article with people's proper names in Galician. *Dialectologia. Special issue* V. 167–190.

Lyons, Christopher. 1999. *Definiteness*. Cambridge: Cambridge University Press.

Mannu, Salvatore. 2017. *Sa lìngua sàrda: Compéndiu de grammàtica normatìva*. Dolianova: Edizioni Grafica del Parteolla.

Marouzeau, Jules. 1922. *L'ordre des mots dans la phrase latine. I: Les groupes nominaux*. Paris: Champion.

Matushansky, Ora. 2006. Why Rose is the Rose: On the use of definite articles in proper names. In Olivier Bonami & Patricia Cabredo Hofherr (eds.), *Empirical issues in syntax and semantics 6 (Papers from CSSP 2005)*, 285–307. http://www.cssp.cnrs.fr/eiss6.

Mikkelsen, Line. 2005. *Copular clauses. Specification, predication and equation*. Amsterdam & Philadelphia: John Benjamins.

Miron-Fulea, Mihaela, Carmen Dobrovie-Sorin & Ion Giurgea. 2013. Proper names. In Carmen Dobrovie-Sorin & Ion Giurgea (eds.), *A reference grammar of Romanian. Volume 1: The noun phrase*, 719–745. Amsterdam & Philadelphia: John Benjamins.

Montero Curiel, Pilar. 2006. *El extremeño*. Madrid: Arco Libros.

Mulder, Walter de & Anne Carlier. 2011. The grammaticalization of definite articles. In Heiko Narrog & Bernd Heine (eds.), *The Oxford handbook of grammaticalization*, 522–534. Oxford: Oxford University Press.

Nübling, Damaris. 2005. Zwischen Syntagmatik und Paradigmatik: Grammatische Eigennamenmarker und ihre Typologie. *Zeitschrift für Germanistische Linguistik* 33(1). 25–56.

Nübling, Damaris. 2012. Auf dem Wege zu Nicht-Flektierbaren: Die Deflexion der deutschen Eigennamen diachron und synchron. In Björn Rothstein (ed.), *Nicht-flektierende Wortarten* (Linguistik – Impulse und Tendenzen 47), 224–246. Berlin & New York: Walter de Gruyter.

Nübling, Damaris. 2017. The growing distance between proper names and common nouns in German: On the way to onymic schema constancy. In Tanja Ackermann & Barbara

Schlücker (eds.), *The morphosyntax of proper names*. [Special issue]. *Folia Linguistica* 51(2). 341–367.

Ortiz Ciscomani, Rosa María & Jeanett Reynoso Noverón. 2012. La determinación y el nombre propio. Un estudio histórico de pragmática social en español. In Emilio Montero Cartelle & Carmen Manzano Rovira (eds.), *Actas del VIII Congreso Internacional de Historia de la Lengua Española*, 2313–2324. Madrid: Arco Libros.

Ortmann, Albert. 2014. Definite article asymmetries and concept types: Semantic and pragmatic uniqueness. In Thomas Gamerschlag, Doris Gerland, Rainer Osswald & Wiebke Petersen (eds.), *Frames and concept types: Applications in language and philosophy* (Studies in linguistics and philosophy 94), 293–321. Cham: Springer.

Pană Dindelegan, Gabriela (ed.). 2013. *The grammar of Romanian*. Oxford: Oxford University Press.

Pană Dindelegan, Gabriela (ed.). 2016. *The syntax of Old Romanian*. Oxford: Oxford University Press.

Perini, Mário Alberto. 2002. *Modern Portuguese: A reference grammar*. New Haven: Yale University Press.

Quesada Pacheco, Miguel Angel. 2002. *El español de América*. Cartago: Editorial Tecnológica de Costa Rica.

Rabella Ribas, Joan Anton. 2006. L'article i el nom propi. In *Homenatge a Joseph Gulsoy* (Estudis de llengua i literatura catalanes 53), 215–230. Barcelona: Abadia de Montserrat.

RAE/ASALE = Real Academia Española & Asociación de Academias de la Lengua (eds.). 2009. *Nueva gramática de la lengua española. Morfología. Sintaxis I. Sintaxis II*. Madrid: Espasa Libros.

Raposo, Eduardo Buzaglo Paiva & Maria Fernanda Bacelar do Nascimento. 2013. Nomes próprios. In Eduardo Buzaglo Paiva Raposo et al. (eds.), *Gramática do Português*, vol. 1, 993–1041. Lisboa: Fundação Calouste Gulbenkian.

Real Academia Española. 1771. *Gramática de la lengua castellana*. Madrid: Ibarra.

Remacle, Louis. 1952. *Syntaxe du parler wallon de La Gleize. Tome I. Noms et articles – Adjectifs et pronoms*. Paris: Les Belles Lettres.

Renzi, Lorenzo, Giampaolo Salvi & Anna Cardinaletti. 2001. *Grande grammatica italiana di consultazione. Volume 1: La frase. I sintagmi nominati e preposizione*. Bologna: Il Mulino.

Riegel, Martin, Jean-Christophe Pellat & Rioul René. 2018. *Grammaire méthodique du français*. 7th edn. Paris: Presses Universitaires de France.

Roehrs, Dorian. 2005. Pronouns are determiners after all. In Marcel den Dikken & Christina Tortora (eds.), *The function of function words and functional categories* (Linguistik Aktuell/Linguistics Today 78), 251–285. Amsterdam & Philadelphia: John Benjamins.

Rohlfs, Gerhard. 1969. *Grammatica storica della lingua italiana e dei suoi dialetti. Sintassi e formazione delle parole*. Torino: Einaudi.

Ronjat, Jules. 1937. *Grammaire istorique des parlers provençaux modernes. Tome III. Deuxième Partie: Morphologie et formation des mots. Troisième partie: Notes de syntaxe*. Montpellier: Société des Langues Romanes.

Rothstein, Susan. 2017. Proper names in Modern Hebrew construct phrases. In Tanja Ackermann & Barbara Schlücker (eds.), *The morphosyntax of proper names*. [Special issue]. *Folia Linguistica* 51(2). 419–451.

Schwarz, Florian. 2009. *Two types of definites in natural language*. Amherst: University of Massachusetts PhD thesis.

Schwarz, Florian. 2013. Two kinds of definites cross-linguistically. *Language and Linguistics Compass* 7(10). 534–559.
Schwarz, Florian. 2019. Weak vs. strong definite articles: Meaning and form across languages. In Ana Aguilar-Guevara, Julia Pozas Loyo & Violeta Vázquez-Rojas Maldonado (eds.), *Definiteness across languages*, 1–37. Berlin: Language Science Press.
Seiler, Guido. 2019. Non-Standard Average European. In Andreas Nievergelt & Ludwig Rübekeil (eds.), *'athe in palice, athe in anderu sumeuuelicheru stedi'. Raum und Sprache. Festschrift für Elvira Glaser zum 65. Geburtstag*, 541–554. Heidelberg: Winter.
Sigurðsson, Halldór Ármann. 2006. The Icelandic noun phrase: Central traits. *Arkiv för nordisk filologi* 121. 193–236.
Sigurðsson, Halldór Ármann & Jim Wood. 2020. "We Olaf": Pro[(x-)NP] constructions in Icelandic and beyond. *Glossa: a journal of general linguistics* 5(1). 16. 1–26.
Silva, Giselle Machline de O. 1996a. Realização facultativa do artigo definido diante de possessivo e de patronímico. In Giselle Machline de O. Silva and Maria Marta Pereira Scherre (eds.), *Padrões sociolingüísticos: análise de fenômenos variáveis do português falado na cidade do Rio de Janeiro*, 119–145. Rio de Janeiro: Tempo Brasileiro.
Silva, Giselle Machline de O. 1996b. O emprego do artigo diante de possessivos e de patronímicos: resultados sociais. In Giselle Machline de O. Silva and Maria Marta Pereira Scherre (eds.), *Padrões sociolingüísticos: análise de fenômenos variáveis do português falado na cidade do Rio de Janeiro*, 265–281. Rio de Janeiro: Tempo Brasileiro.
Sousa Fernández, Xulio. 1994. O artigo cos nomes propios de persoa no galego moderno. In Ramón Lorenzo Vázquez (ed.), *Actas do XIX Congreso Internacional de Lingüística e Filoloxía Románicas. Universidade de Santiago de Compostela, 1989*, 309–316. A Coruña: Fundación Pedro Barrié de la Maza, Conde de Fenosa.
Spescha, Arnold. 1989. *Grammatica sursilvana*. Cuera: Casa Editura per Mieds d'Instrucziun.
Stolz, Thomas and Nataliya Levkovych. this volume. On *Special Onymic Grammar (SOG)*: Definiteness markers in Fijian and selected Austronesian languages.
Sturm, Afra. 2005. *Eigennamen und Definitheit* (Linguistische Arbeiten 498). Tübingen: Niemeyer.
Szczepaniak, Renata. 2009. *Grammatikalisierung im Deutschen. Eine Einführung*. Tübingen: Narr.
Szczepaniak, Renata & Johanna Flick. 2020. Introduction. In Renata Szczepaniak & Johanna Flick (eds.), *Walking on the grammaticalization path of the definite article: Functional main and side roads* (Studies in Language Variation 23), 1–13. Amsterdam & Philadelphia: John Benjamins.
Thomas, Earl W. 1969. *The syntax of spoken Brazilian Portuguese*. Nashville: Vanderbilt University Press.
Toscano Mateus, Humberto. 1953. *El español en el Ecuador*. Madrid: Escelicer.
Van de Velde, Mark. 2019. Nominal morphology and syntax. In Mark Van de Velde, Koen Bostoen, Derek Nurse & Gérard Philippson (eds), *The Bantu languages*, 237–269. 2nd edn. London & New York: Routledge.
Vidal de Battini, Berta Elena. 1949. *El habla rural de San Luis. Parte I: Fonética, morfología, sintaxis*. Buenos Aires: Universidad de Buenos Aires.
Werth, Alexander. 2020. *Morphosyntax und Pragmatik in Konkurrenz. Der Definitartikel bei Personennamen in den regionalen und historischen Varietäten des Deutschen* (Studia Linguistica Germanica 136). Berlin & Boston: Walter de Gruyter.

Appendix 1

The following table contains the sample of languages and language varieties that constitutes the data base for the synchronic and diachronic investigation of the occurrence of the definite article with personal names presented above.

Table 6: Language sample.

n	Language	References
1.	Asturian	ALLA (2001)
2.	Catalan	ALDC, GEIEC (2018)
3.	French	Grevisse and Goose (2016), Riegel et al. (2018)
4.	Galician	ALGa, Álvarez and Xove (2002), Louredo Rodríguez (2015), Sousa Fernández (1994)
5.	Italian	Kubo (2016), Renzi et al. (2001), Rohlfs (1969)
6.	European Portuguese	Cunha and Cintra (2017), Raposo and do Nascimento (2013)
7.	Brazilian Portuguese	Perini (2002), Thomas (1969)
8.	Occitan	ALG, Joly (1971), Ronjat (1937)
9.	Romanian	AR (2008), Dobrovie-Sorin and Giurgea (2013), Pană Dindelegan (2013, 2016)
10.	Sardinian	Blasco Ferrer (1986; 2003), Jones (2003), Mannu (2017)
11.	Spanish	Fernández Leborans (1999), RAE/ASALE (2009)
12.	Sursilvan	Spescha (1989)

Appendix 2

The following table provides an overview of the definite article forms in the surveyed Romance languages. In Romance languages, the definite article occurs before the noun, with the exception of Romanian, where the definite article is attached to the noun. The genitive-dative forms of the definite article in Romanian are not included in the table. The article forms occurring before vowels are indicated in parentheses.

Table 7: Definite article forms in the Romance languages surveyed.

Language	Masculine	Feminine
Asturian	el	la
Balearic Catalan	es, el (l'), en (n')	sa, la (l'), na (n')
Central Catalan	el (l'), en (l')	la (l')
French	le (l')	la (l')
Galician	o	a
Italian	il, lo (l')	la (l')
Occitan	lou (l')	la (l')
Portuguese	o	a
Romanian	-(u)l, -le, -a	-a
Sardinian	su (s')	sa (s')
Spanish	el	la
Sursilvan	il (igl, gl')	la (l')

Yves D'hulst, Rolf Thieroff and Trudel Meisenburg
River names
Definite articles and place names in West-Germanic and Romance

Abstract: Germanic and Romance languages with prenominal determiners always require the use of the definite article with river names. This property sets river names apart from other place names that either do not take an article (city names) or display strong cross-linguistic variation (country names). We argue that the requirement to appear with a definite article correlates with a fundamental semantic property that distinguishes river names and common nouns that refer to rivers from other proper names and common nouns: their lexical underspecification for boundedness. We argue that this underspecification must be resolved syntactically through the presence of a definiteness feature provided by the definite article.

Keywords: definiteness, determiners, proper names, river names, (un)boundedness.

1 Introduction

The use of determiners greatly varies both within languages and across language families. Part of this variation depends on the determiner's morphological complexity in terms of number of semantic and grammatical features (Baerman et al. 2017). German *das* 'the$_{\text{NOM.N.SG}}$' expresses definiteness, possibly including specificity, gender (neuter), number (singular) and case (nominative or accusative). We argue that each of these features may trigger the occurrence of the article. On the other hand, English *the* has no grammatical markings, hence only expresses definiteness (and specificity) and as a consequence can only be triggered by this semantic feature.

Proper names referring to geographical entities (henceforth place names, or, more specifically, city, river etc. names) display particularly interesting patterns: city names usually reject the determiner in Germanic and Romance languages; country names behave similarly in most Germanic and Ibero-Romance languages, vary according to gender in German, take the definite article with most but not all country names in Portuguese and obligatorily take the article in all other Romance languages; river names take an article in all West-Germanic and Romance languages with prenominal determiners (see Caro Reina 2020 for a more in-depth assessment of the Romance languages).

The varying use of the definite article with place names is surprising, since proper names are usually assumed to be inherently definite and therefore do not require the expression of an additional morpheme carrying definiteness (the exceptional use of definite determiners with personal names will be addressed in Section 2). The behaviour of place names is particularly striking in the case of river names, and our central concern is to understand why these expressions are used with a definite article in all Germanic and Romance languages with prenominal determiners. This question falls into two components: (i) what is the function of the definite article with river names? and (ii) what is the ontological property – if any – that triggers this function? English, where the article lacks grammatical features, helps to solve the first question: since the English definite article is only endowed with a definiteness feature, it is the general function of this feature that must be involved with river names. This brings river names more in line with common nouns than any other sort of proper name. The simple fact that river names all behave the same in the languages under examination shows that the ontological property of river names has a strong linguistic underpinning. We will show that river names and common nouns referring to rivers share important semantic properties, and thus may be said to constitute a natural subclass of nominal expressions.

We will study the use of the definite article with river names in a set of languages that display a fairly homogeneous behaviour in other contexts as well, namely West-Germanic (especially English, Dutch and German) and Romance languages, with the exception of Romanian. These languages share the property of having a prenominal definite article and the lack of an article in front of city names. These properties constitute the anchor points to assess river names. We hope to be able to integrate the remaining Germanic and Romance languages in later research. A similar caveat holds for the inclusion of a diachronic perspective. Although we acknowledge the relevance and interest of diachronic data, we designed our study in a strict synchronic fashion. It goes without doubt that in earlier stages of the languages under scrutiny the definite article was lacking with river names, just as it did with other place names that nowadays require the article. But in order for such diachronic data to be really revealing, a comparative chronology of the use of the definite article in front of place names (at least country, city and river names) must be modelled.¹ Such a challenge goes beyond the aim of this paper, and we leave it for further research.

This paper is organised as follows. In Section 2 we present our main theoretical assumptions on the syntax and semantics of nouns. River names are positioned within a rough semantic and distributional typology of place names in Section 3. In Section 4 we show that definiteness must be disregarded as trigger for the realization of a definite article with some place names but not so with river names and Section 5 reveals semantic properties that characterize river names, but not other proper names or common nouns. We argue in Section 6 that these properties most likely follow from the boundedness feature associated to nominal predicates: the boundedness feature remains lexically underspecified with nominal expressions referring to rivers and this underspecification is syntactically resolved by the definite article.

2 Theoretical background

The syntax and semantics of the noun phrase have been the subject of intensive linguistic research, inspiring generations of linguists to unravel the intricate relationships between determiners and nouns and between linguistic expressions and the referential world (Russell 1905; Carlson 1977, 1980; Heim 1982; Löbner 1985; Enç 1991; Diesing 1992; Longobardi 1994; Chierchia 1998, 2003; among many others). These relationships are heavily codetermined by what one assumes nouns to be. The views roughly fall into two veins. Following the classical grammatical tradition, some consider nouns as inherently entity-denoting and thus classify them as referential expressions of type <e> (see Chierchia 1998; Baker 2003). The alternative stance takes nouns as property-denoting expressions and identifies them as predicates of type <e,t> (see Higginbotham 1985, 1987; Longobardi 1994 and later work; McNally 1995, 2004; Dobrovie-Sorin and

---

1 The development of the demonstrative determiner into the definite article in the Romance languages has been thoroughly covered in the literature (for an overview, see Vincent 1997). The emergence of the article in specific syntactic contexts (common nouns vs. proper names and the broad typology of proper names) has not been covered as accurately (however, see Renzi 1976, 1979). As for West-Germanic languages, German onomastics is a welcome exception and recent years have witnessed an increasing interest in the (syntactic) variation in the use of definite determiners with proper names (among others: Werth 2014; Ackermann 2018; Nübling 2020; Schmuck 2020a, 2020b).

Giurgea 2015); they convert into entity-denoting expressions through the interaction with determiners of various sorts.

Although the two views fundamentally differ in their definition of nouns, there is only a fine line between the arguments in favour of the one or the other view. Eventually both require some type-shifting operation in order to accommodate either the property-denoting interpretation of the bracketed expression in (1a) or the entity-denoting interpretation of the nominal phrase in (1b).

(1) English
    a. *She called her son [a liar / John].*
    b. *She called [her son / John].*

In this paper we will espouse the view that nouns, including proper names, are basically predicates, i.e. expressions of type <e,t>. The main reason for doing so, is that this semantic view interacts in interesting ways with the assumptions about the syntactic relationships between determiner and noun, commonly referred to as the "DP-hypothesis".[2]

The DP-hypothesis emerged in the eighties (Brame 1982; Fukui and Speas 1986; Abney 1987) and was further developed over several decades (Szabolcsi 1983 and later work; Giorgi and Longobardi 1991; Giusti 1991 and later work; Longobardi 1994 and later work; among many others). It takes (argumental) NPs to be embedded within DPs, as sketched out in (2). It is fundamentally a hypothesis about the syntactic structure of noun phrases and does not exclude *per se* one or the other view on the semantics of nouns. Nevertheless, the fact that referential noun phrases in argument positions mostly require the presence of a determiner in Romance languages (3a), and the fact that non-referential noun phrases in predicative positions are compatible with determinerless nouns (3b) rather suggest that nouns are genuinely predicative and convert into referential expressions through the determiner.

---

2 In this paper we will adopt the framework of the DP-hypothesis as a descriptive tool: the DP-hypothesis has proven to accurately relate several aspects of nominal syntax within the language families that we are studying here. At a theoretical level, the DP-hypothesis remains controversial (see the recent debate in *Glossa*: Bruening 2020 vs. Salzmann 2020).

(2)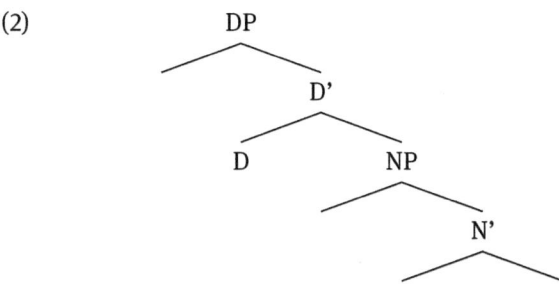

(3) Italian
    a. *Ho comprato *(del) vino.*
       have.PRS.1SG bought wine
       'I have bought wine.'
    b. *Gianni è ingegnere.*
       John be.PRS.3SG engineer
       'John is an engineer.'

Within the DP-hypothesis and following Longobardi (1994), it is assumed that (personal) proper names move to D. The relative order of the proper name and the possessive clearly shows that this movement takes place in overt syntax in Italian. Possessive adjectives occur between article and common noun whenever they are not focused or contrasted (4a); however, with personal names they occur post-nominally. The lack of determiner, the post-nominal position of the possessive adjective and its non-focused or non-contrastive reading can all be accounted for under the single assumption that the proper name moves to D, as shown in (4b).

(4) Italian
    a. *il mio cappello*
       the my hat
       'my hat'
    b. *Gianni mio*
       'my Gianni'

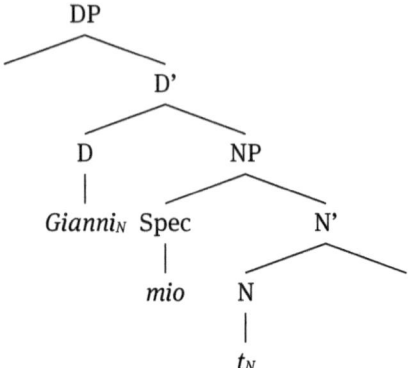

Proper names are known to differ with respect to the way they are affected by N-to-D movement. While N-to-D movement is overt in Romance, Longobardi (1994) has argued that it remains covert in English, witness the A-N order of the English example in (5) vs. the N-A order of Italian in (6a). Furthermore, there are varieties that optionally allow for procrastination of N-to-D movement and insert an arguably expletive determiner, mostly for pragmatic reasons (familiarity), both in West-Germanic, e.g. German (7), and Romance, e.g. Italian (6b).[3]

(5)   English [Longobardi 1994: 628]
              Old John

(6)   Italian
    a.   *Cameresi vecchio* [Longobardi 1994: 624]
         Cameresi old
         'Cameresi the elder'
    b.   *La Maria arriva domani.*
         the Mary arrives tomorrow
         'Mary is coming tomorrow.'

---

3 In some Romance languages personal names combine with a definite article in a systematic way. This is the case in Portuguese and Catalan, irrespective of gender; other varieties require a definite article only in the feminine (e.g. Romanian). As pointed out by an anonymous reviewer, Catalan even developed a specific onymic definite article (Caro Reina 2014). For a detailed analysis of the use of definite articles with personal names, see Caro Reina (this volume). Both in German and in Dutch the use of the definite article with personal names is subject to diatopic variation (Nübling et al. 2015: Ch. 7).

(7)   German
      Der Peter kommt morgen.
      the Peter comes tomorrow
      'Peter is coming tomorrow.'

N-to-D movement divides languages into two broad groupings (Longobardi 2003, 2005, 2008). In West-Germanic and all Romance languages but Romanian, N-to-D movement is limited to most but not all proper names, typically personal names and city names (8a), and optionally a handful of proper nouns (typically those expressing the concept of 'house' (8b); Longobardi 2001). In these languages, the definite article is prenominal (8c) and when N-to-D movement takes place, no determiner is realized (8b). North-Germanic and Romanian belong to the other grouping, where movement to the D-domain, by either noun or adjective (9b–c), is generalized for definite descriptions. When such movement occurs, the noun or the adjective combines with the definite article. In these languages definite articles are enclitic.

(8)   Italian
  a   *Maria, Roma*
      Mary, Rome
  b   *casa mia*
      house my
      'my house'
  c   *la mia casa*
      the my house
      'my house'

(9)   Romanian
  a   *Mari-a, Bucureşti(-ul)*
      Mary-the, Bucharest(-the)
  b   *palat-ul frumos*
      palace-the beautiful
      'the beautiful palace'
  c   *frumos-ul palat*
      beautiful-the palace
      'the beautiful palace'

In this paper we will only study the use of the definite determiner with river names in languages belonging to the first grouping. North-Germanic and Romanian have been omitted because N-to-D movement does not exclude the realization of the determiner with common nouns and the behaviour of proper names with respect to the realization of the definite article is inconsistent across these languages (North-Germanic vs. Romanian) and within Romanian itself: North-Germanic does not realize a definite article with (unmodified) personal names, and Romanian realizes the definite article only with a subset of feminine personal names (those that derive from first declension class Latin personal names) and optionally with city names (9a).[4]

3 Place names

Not all proper names behave the same with respect to N-to-D movement and the realization of the definite article. Especially among place names, there is a huge variation, both language-internally and cross-linguistically.[5] Within the selection of languages that we will be considering in this paper, namely West-Germanic and Romance languages with prenominal definite article, city names generally appear without determiner (10a)–(11a). Just as with personal names, they typically trigger covert N-to-D movement in West-Germanic, witness the A-

---

4 As for river names, the definite article is obligatory in the only Romance language with a postnominal determiner, Romanian (*Rinul este un fluviu în Europa* 'Rhine$_{DEF}$ is a river in Europe'); in the Scandinavian languages river names are in general used without a definite article (Icelandic *Rín er þriðja lengsta fljót Evrópu* 'Rhine is third longest river Europe$_{GEN}$'; Swedish *Rhen är Västeuropas största flod* 'Rhine is Western Europe's longest river'; Norwegian *Donau er Europas nest lengste elv, etter Volga* 'Danube is Europe's second longest river, after Volga'; Danish *Donau er den næstlængste flod i Europa, efter Volga* 'Danube is the second longest river in Europe, after Volga'). In both Norwegian and Danish, a few river names end in *-en* (*Elben, Rhinen, Seinen, Slien, Themsen, Tiberen* and *Nilen*). The suffix *-en* is the non-neuter definite article, but there is some evidence that in modern Danish and Norwegian *-en* with river names is re-interpreted as part of the river name, not as the definite article. On the other hand, in Danish the degrammaticalization of *-en* is not yet fully accomplished. Thus, with an adjectival attribute, neither the form without *-en* seems to be possible (**den tyske Rhin* 'the German Rhine'), nor the form with *-en* is available (**den tyske Rhinen* 'the German Rhine.en'). The first should be possible if *-en* were the definite article (which is omitted in Danish when the prenominal definite article (*den*) with a preposed attribute is used), the second would be employed if *-en* were part of the noun; instead Danish uses *den tyske flod Rhinen* 'the German river Rhine'.
5 Throughout this paper, we will only be concerned with unmodified place names in referential contexts. Place names that are modified by adjectives (*the blue Danube*), relative clauses (*the London that never was*) and the like obey different conditions and mostly require the realization of an overt determiner.

N order of English (10b) (compare with (5)) and overt N-to-D movement in Romance, witness the N-A order of Italian (11b) (compare with (6a)).

(10) West-Germanic
  a. En. *(*the) London*, Du. *(*het) Brussel*, Ge. *(*das) Berlin*
  b. *Old Rome*

(11) Romance
  a. Sp. *(*el) Madrid*, It. *(*la) Roma*, *(*le) Paris*
  b. *Roma antica* [Longobardi 1994: 624]
     Rome antique
     'Old Rome'

River names, on the other hand, obligatorily occur with a definite article. As has been pointed out on various occasions, the English definite article cannot be used expletively (among others: Vergnaud and Zubizarreta 1992; Longobardi 1994). We may therefore assume that its status is not to be equated with the definite article in (7).

(12) West-Germanic
  En.*(the) Thames*, Du. **(de) Schelde*, Fs.**(de) Geau*, Ge.**(der) Rhein*

(13) Romance
  Sp.**(el) Tajo*, Pt. **(o) Douro*, It.**(il) Po*, Fr.**(la) Seine*

The use of definite articles with country names is very diverse and ranges from impossible in western Germanic (14) to obligatory in Gallo-Italo Romance (18). Several languages display an intermediate status: German (15) uses the definite article optionally with masculine and obligatorily with feminine but never with neuter country names (see Thieroff 2000), Spanish (16) tends to use country names without definite article, but optionally inserts one with some country names, Portuguese (17) optionally takes the definite article with most country names.[6]

---

[6] Contrary to fact, the examples in (15)–(17) might suggest that German and especially Spanish and Portuguese behave similarly with respect to the use of the definite article with country names. In German the overwhelming majority of country names are neuter and hence do not combine with the definite article. Spanish is similar, in the sense that most country names do not take the definite article. However, as an anonymous reviewer points out, there are Spanish

(14) West-Germanic (with the exception of German)
En. *(*the) England*, Du. *(*het) België* 'Belgium'

(15) German
(*das) Deutschland, (der) Irak, *(die) Schweiz
'Germany, Iraq, Switzerland'

(16) Spanish
(*la) España, (la) Argentina
'Spain, Argentina'

(17) Portuguese
(o) Brasil, (*o) Portugal
'Brazil, Portugal'

(18) Gallo-Italo Romance
Fr.**(la) France , *(le) Danemark*; It.**(l')Italia, *(il) Messico*
'France, Denmark; Italy, Mexico'

The typology of place names, of course, is not limited to city, country and river names, but includes a large amount of often fine-grained distinctions (regions and continents, hills and mountains, streets, squares and highways, lakes, seas and oceans,[7] planets and stars, and many more; a more systematic taxonomy of proper names is to be found in Nübling et al. 2015: Ch. 9 and Van Langendonck 2007: 202–218). We will not dig into these distinctions, but limit ourselves to two broad observations. First, the grammar of these other place names, i.e. their interaction with determiners, varies from language to language (and often even within a single language), but conforms to one of the three types presented above. Second, the fact that there is some correlation between the grammar and the typology of place names shows that the onomasiological orientation of this typology has indeed linguistic relevance, in spite of the fact that the typology of common nouns is more semasiological (*count – collective – mass*).

---

varieties that generalize the definite article with country names and thus resemble the Gallo-Italo Romance setting. In Portuguese most country names appear with a definite article.

[7] The status of lakes, seas and oceans as proper names is remarkable, since they mostly must combine with a common noun (*lake, sea, ocean* etc.). Harweg (1983) therefore qualifies these place names as 'appellative proper names'.

4 Definite articles and syncretism

In spite of many controversies concerning the exact formulation of the semantic and pragmatic value of definite articles and even stronger controversies concerning their use and interpretation and the number of their functions, there is a broad consensus that the definite article mainly serves to introduce a unique entity-referring noun phrase whose referent is known (and prominent) to both speaker and hearer.[8] This function is illustrated in (19). We will identify it as the definiteness feature ([def]) associated to the definite article.

(19)   German
        *Ich habe gestern ein Buch und eine Zeitung gekauft, und jetzt weiß ich nicht mehr, wo ich **das** Buch hingelegt habe.*
        'I've bought a book and a newspaper yesterday, and now I can't remember where I put **the** book.'

In most West-Germanic and Romance languages definite articles are not only endowed with a definiteness feature, but with several others, notably gender and number features (especially in Romance languages) and sometimes also a case feature (German). In most studies on determiners, it is tacitly assumed that whenever a definite article is used, all these features are equally relevant: since the features are syncretically realized, no feature can be ignored. We challenge this tacit assumption and argue (i) that the features associated to the definite article are highly structured and (ii) that this structured hierarchy may affect the active status of features.

Consider in this respect the variation in German in (15). There is no plausible cause that could explain why the definiteness feature should be more relevant with feminine country names than with masculine or neuter country names. We therefore conclude that, since the [def] feature is irrelevant for neuter country names, it is irrelevant for all country names: in this respect German patterns with English and not with most Romance languages (where country names do not move to D and as a consequence require an expletive determiner to avoid a default existential interpretation; see Longobardi 1994). The natural corollary of this conclusion is that the obligatory use of the definite article with feminine and the optional use with masculine country names is solely related to

---

8 This qualification draws on Russell's (1905) original definition. See Löbner (1985) for a thorough discussion.

gender and, crucially, that there must be a fundamental difference in the grammatical representation of the gender values (masculine and feminine in Romance, masculine, feminine and neuter or neuter and non-neuter in Germanic).[9]

The appearance of the definite article in German can be captured under a feature-geometric approach. The central claims in this approach are that features are privative, in Trubetzkoy's (1939) sense, and strictly hierarchically structured along feature nodes (such as the GENDER node in (20)). Feature geometry was first proposed by Clements (1985) in his study on phonological features. Harley and Ritter (2002a, 2002b) have shown how the approach can be fruitfully applied in the case of morpho-semantic features (for a psycholinguistic evaluation of morpho-syntactic feature geometries, see Fuchs et al. 2015). A feature-geometrical analysis of a three-way gender distinction like the German one (and more generally the Germanic one, with English as the exception) is outlined in (20):[10] neuter is unmarked for gender; masculine is underspecified and feminine is fully specified for gender. Consequentially, no definite article may be used with neuter country names, because country names do not require a definiteness feature and because neuter is featureless. On the other hand, the definite article is obligatory for feminine nouns, which are fully specified for gender, and optional for underspecified masculine country names.

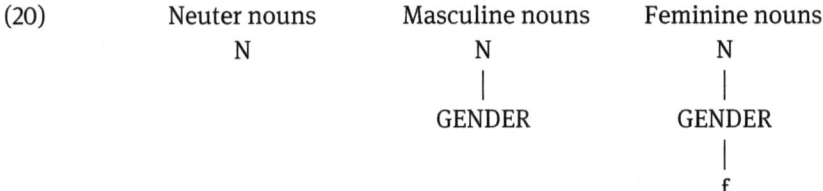

(20) Neuter nouns Masculine nouns Feminine nouns
       N   N   N
                      | |
                      GENDER   GENDER
                                       |
                                       f

By contrast, Italian and French realize a definite article with all country names (18). It is not *per se* excluded that Italian and French realize the definite article, by need to express the definiteness feature with country names, but it is rather unlikely: after all a principled reason why this should be the case for Italian and

---

9 The use of the definite article with plural place names is to be accounted for in a similar fashion: under the standard assumption that plural is a marked feature (most explicitly advocated by Harley and Ritter 2002a, 2002b) this feature needs to be overtly expressed with plural place names, thus triggering the presence of the definite article.

10 In this paper, we are only concerned with the grammatical notion of gender. Gender is to be distinguished from natural sex, the notion relevant for animate entities that strongly interacts with gender (among others: Corbett 1991, 2015; Harris 1991; and especially for German: Nübling 2014; Nübling and Fahlbusch 2014, 2016).

French (18) and not for western Germanic (14) is hard to identify. The syntactic and semantic conditions bearing on empty D heads in Romance, first observed and identified by Longobardi (1994) and briefly summarized in Section 2, provide a fruitful alternative: in order to avoid the application of the default existential interpretation of the Italian and French country names, an expletive determiner is inserted. The definite article is merely a placeholder and does not directly contribute to the semantic computation of the noun phrase. Spanish and Portuguese provide clues that this analysis is on the right track: the optional use of definite articles in (16)–(17) is only available when they may qualify as expletive (see Longobardi 1994); and it is only then that they may serve mostly vague pragmatic purposes as is the case with personal names (see (6b) in Italian or (7) in German).

5 Rigidity vs. referential vagueness

The fact that the definite article occurs with river names in all Germanic and Romance languages with prenominal determiners strongly suggests that its appearance is directly linked to its definiteness feature. Other features such as gender, number and case can immediately be discarded, since the English definite article has no marking for any of these features and still appears with river names. The conditions governing the appearance of the definite article with Romance country names are equally irrelevant, since such conditions would wrongly predict a contrast between West-Germanic and Romance languages.

It thus seems that river names are genuinely different from other place names and perhaps more similar to common nouns. This view receives support from the contrast between the two sentences in (21). (21a) can only be uttered when France is generally affected by strikes, not just the part of France round Paris or any other random part of France. (21b) is radically different: the Seine probably just bursts its banks in the neighbourhood of Paris or some other (presumably upstream) part of the Seine (that may affect the level in Paris), not necessarily along its over 700 km length.

(21)   English
   a.   *Perhaps it is not such a good idea to visit Paris while France is on strike.*
   b.   *Perhaps it is not such a good idea to visit Paris while the Seine bursts its banks.*

What river names seem to share with common nouns is the fact that, unlike other proper names, they are not rigid designators (see Berezowski 1997). The term 'rigid designator' stems from Kripke (1980) and refers to expressions designating the same entity in all possible worlds this entity exists in. It is commonly assumed that proper names are rigid designators and common nouns are not (see LaPorte 2016 for discussion). If we look at the examples in (22), however, we notice that the same expression designates distinct entities.[11] In (22a) the entity referred to by the expression *Colorado* most likely implies the notion of water, while in (22b) it definitely does not. Similarly, in (22c) reference is made to the length of the course of the river, while in (22d) the water referred to does not even lie within the course of the Amazon.

(22) English
   a. *I dreamt I fell in the Colorado.*
   b. *It was unreal to just stand in the Colorado, after months of extreme drought.*
   c. *The Amazon is 6.992 km long.*
   d. *The Amazon colours the Atlantic grey.*

The kind of referential vagueness observed in (22) distinguishes river names from country or city names, and more generally from proper names, and sets them on a par with common nouns: in (23a) the expression *médecin* 'doctor' may either refer to a qualification, or to a profession or to both. Under our view, it is the function of the determiner to bind the lexical predicative property that is relevant in discourse, as shown by the interpretive difference with (23b), whose reading implies the professional qualification (see D'hulst 2007). In the case of river names, the definite article performs this very same function, singling out the relevant predicative property associated with the proper name.

---

11 The notion of rigidity adopted here slightly differs from the received one. Commonly rigidity is intended to capture the non-disjointness of referents (LaPorte 2016): an expression such as *John Fitzgerald Kennedy* will under a common worldview always refer to the individual that held the office of President of the United States from 1961 until 1963. On the other hand, the expression *President of the United States* may or may not refer to the individual whose name was John Fitzgerald Kennedy. When it does not refer to that individual the expressions *President of the United States* and John Fitzgerald Kennedy have disjoint referents. It may be said that in the case of river names, the reduced or even lack of rigidity does not necessarily involve disjointness: under the natural reading of (21b) where *The Seine* refers to the Parisian section of the river, the referent of *The Seine* is not disjoint from other sections of the river we know as the Seine. Non-disjointness, however, does not imply identity.

(23) French
- a. *Jean est   un médecin.*
 John be.PRS.3SG a   doctor
 'John is a doctor.'
- b. *Jean est   médecin.*
 John be.PRS.3SG doctor
 'John is a doctor.'

One could of course argue that the referential vagueness illustrated in (22) is nothing but the effect of metonymic use of river names and therefore not fundamentally different from what can be observed with other place names or any other proper or common noun. However, this conclusion is not warranted. Standard cases of metonymy, like those in (24) and (25), do not allow for coordination when distinct entities are referred to in the coordinated sentences (24c)–(25c). This holds both for common nouns (24) and proper names (25). However, river names behave differently: even when they clearly refer to distinct river parts, they can be coordinated as in (25d).[12]

(24) French
- a. *Le tribunal a été construit en l'an 1913.*
 'The court has been built in the year 1913.'
- b. *Le Tribunal est une institution des Nations Unies.*
 'The Court is an institution of the United Nations.'
- c. **Le tribunal a été construit en l'an 1913 et est une institution des Nations Unies.*

---

[12] River names may be used to denote the states or regions that they flow through. This may be considered as metonymy. In such examples, coordination is excluded as with other proper and common names (see (24c)–(25c)).

(i) French
- a. *Le Mississippi est habité par environ 3 millions d'habitants.*
 'Mississippi is inhabited by about 3 million people.'
- b. *Le Mississippi est le fleuve le plus long de l'Amérique du Nord.*
 'The Mississippi is the longest river of North America.'
- c. **Le Mississippi est habité par environ 3 millions d'habitants et le fleuve le plus long de l'Amérique du Nord.*
 'Mississippi is inhabited by about 3 million people and the longest river of North America.'

'The court has been built in the year 1913 and is an institution of the United Nations.'

(25)   German
    a.   *Karlsruhe ist die zweitgrößte Stadt Baden-Württembergs.*
        'Karlsruhe is the second largest town in Baden-Württemberg.'
    b.   *Karlsruhe entscheidet auch bei einem Streit zwischen Staatsorganen.*
        'Karlsruhe also decides in a conflict between state authorities.'
    c.   **Karlsruhe ist die zweitgrößte Stadt Baden-Württembergs und entscheidet auch bei einem Streit zwischen Staatsorganen.*[13]
        'Karlsruhe is the second largest town in Baden-Württemberg and also decides in a conflict between state authorities.'
    d.   *Die Donau entsteht aus dem Zusammenfluss von Brigach und Breg, ist 2.857 km lang und stellt eine der ältesten und bedeutendsten Handelsrouten Europas dar.*
        'The Danube is formed by the confluence of Brigach and Breg, is 2,857 km long and constitutes one of the oldest and most important trade routes of Europe.'

The examples in (24)–(25) clearly expose the difference between river names on the one hand and other proper names and common nouns on the other. It must be pointed out, however, that the special behaviour of river names also extends to the corresponding common nouns, both with respect to the vagueness of reference (26) as with respect to coordination of conjuncts with distinct referents. In the examples in (26)–(29) the common noun most likely refers to parts of the river and not to the river as a whole.

(26)   English
        The river burst its banks. (cf. (21b))

(27)   German
        Der Fluss ist zugefroren.
        'The river is frozen.'

(28)   Italian

---

13 An anonymous reviewer points out that the ungrammaticality of (25c) may be related to the difference in thematic roles. Note that such differences generally do not play a role in coordination (e.g. *John was ill and went to the doctor*) and that (25d) also coordinates distinct thematic roles (Theme and Agent), without triggering any grammaticality affect.

> *Il fiume è navigabile.*
> 'The river is navigable.'

(29) French

> *Le fleuve est pollué aux PCB.*
> 'The river is polluted by PCB.'

The example in (30) is an exact copy of (25d) with the common noun replacing the river name. It shows that common nouns that refer to rivers or river parts may be coordinated even if the exact entity referred to (source, course, route) changes between conjuncts.

(30) German

> *Der Fluss entsteht aus dem Zusammenfluss von Brigach und Breg, ist 2.857 km lang und stellt eine der ältesten und bedeutendsten Handelsrouten Europas dar.*
> 'The river is formed by the confluence of Brigach and Breg, is 2,857 km long and constitutes one of the oldest and most important trade routes of Europe.'

In this section we have argued that the use of the definite article with river names is inherently related to its definiteness feature, that river names are not rigid designators in the way other proper names are and that they behave in many ways like common nouns or at least as common nouns that take rivers as referents. In other words, expressions that refer to rivers – be they proper names or common nouns – display a syntactic and semantic behaviour that is distinct from other nominal expressions. In the next section we will try to pinpoint the origin of this special behaviour, comparing expressions that refer to rivers (proper and common nouns) with other classes of nouns.

6 Boundedness

Standard categorizations divide common nouns into count and mass nouns. Jackendoff (1991) has a more fine-grained classification that categorises them into four classes: individuals, groups, substances and aggregates. This classification is governed by two general conceptual features: [±b(ounded)] and [±i(nternal structure)]. The boundedness feature discriminates between entities "whose boundaries are [or are not] in view or [...] of concern" (Jackendoff 1991: 19). It sets individuals and groups apart from substances and aggregates.

Groups and aggregates differ from individuals and substances to the extent that only the former comprise a multiplicity of distinguishable individuals and therefore can be said to have an internal structure. The two features allow for the four-way nominal distinction in (31) (Saeed 2009: 284). In the literature, this typology is mostly discussed in reference to common nouns, but there is no principled reason why it should be confined in such a way, as shown in the last line of (31).

(31)   individuals   groups   substances   aggregates
       [+b] [−i]   [+b] [+i]   [−b] [−i]   [−b] [+i]
       a pig   *a committee*   *water*   *buses, cattle*
       Donald   *the Azores*   *Pepsi*   *Ytong*

Nominal expressions referring to rivers resist a categorization along the lines in (31): they intuitively do not qualify as groups, substances or aggregates, nor do they qualify as true individuals. One tends to categorize nouns referring to rivers as expressions that are [+bounded] because we conceive of rivers as bounded by their banks, source and mouth. But in (21b), likewise the corresponding example in (26), it is precisely the boundedness by the banks that disappears. The examples in (27)–(29) can be analysed in a similar manner: the fact that we can access parts of the river follows from the fact that the longitudinal boundedness can be easily ignored (also recall the contrast in (21)).

The intermediate conclusion that nominal expressions referring to rivers are not [+bounded] expressions, like true individuals, does not imply that they are [−bounded]: if they were, then nominal expressions referring to rivers would qualify as substances ([−b], [−i]) which they are clearly not. We therefore assume that they are underspecified for boundedness ([αb]), and characterize them as in (32).

(32)   rivers
       [αb] [−i]
       river, creek
       Thames

We may now address the two final questions: (i) why are nouns referring to rivers underspecified for boundedness ([αb]) and (ii) how does the definite article interact with this property?

As for the first question, nouns referring to rivers have two related semantic properties that distinguish them from other nouns. Their most striking property is that they refer to a substance that flows, typically flowing water.[14] The reference to a flowing substance also indirectly determines their second property: they make reference to a mass (of flowing water) that is hardly quantifiable, precisely because it is a flowing substance.[15]

---

14 This property is also the defining lexical attribute identified in dictionary entries, irrespective of language, as illustrated in the following English and German definitions:

English
>RIVER: n. 1 a large natural **flow** of water along a channel to the sea, a lake, or another river. 2 a large quantity of a **flowing** substance.
>(*Concise Oxford English Dictionary*, Pearsall 2002; boldface ours)

German
>FLUSS, [...] 1. größeres **fließendes** Wasser [...] 2. stetige, **fließende Bewegung**, ununterbrochener Fortgang: der Fluss der Rede, des Straßenverkehrs. Zus.: Gedankenfluss, Redefluss, Verkehrsfluss [...]
>(*Duden. Bedeutungswörterbuch* [4]2010; boldface ours)

15 We thank an anonymous reviewer for pointing out that there might be other proper names that include the notion of a flowing substance, such as waterfalls, hurricanes or winds. We can only tentatively address these categories here.

When they appear in referential argument positions, most waterfalls seem to take the definite article in the languages that we have examined in this paper. However, we are reluctant to take waterfalls as further evidence for the analysis of river names. At a formal level, most names of waterfalls contain a common noun (*fall(s)*, *cascade* and the like) and display some kind of possessive syntax. This renders their morphosyntax more complex than traditional river names (although we do not take the presence of a common noun to automatically trigger the presence of a definite article, as evidenced by English lake names; see below). The complexity of waterfall names also carries over to the heterogeneity of the true onymic part which may or may not take its origin in a place name: the river that the waterfall is part of (e.g. *the Niagara Falls*), or the locality where the waterfall is situated, e.g. *the Aysgarth Falls*, or any arbitrary naming, e.g. *the Victoria Falls*). Semantically, it is not entirely clear whether waterfall names just refer to the waterfall's geographical location or to the moving water. We tend to favor the first view.

In contrast, hurricanes perform quite homogeneously and appear without exception without definite article. One must note, however, that the naming of hurricanes follows a strict naming convention defined by the World Meteorological Organization (https://public.wmo.int/en/our-mandate/focus-areas/natural-hazards-and-disaster-risk-reduction/tropical-cyclones/Naming). This convention requires hurricanes to receive a personal name. Our calculated guess is that the syntax of personal names is carried over to hurricanes. Winds (*mistral, bora, sirocco* etc.), on the other hand, qualify as 'category-designating proper names' (see Nübling this volume). They take the definite article in the languages we have studied, although there is some fluctuation with some names (e.g. *Khamsin*) in English.

These properties distinguish nouns referring to rivers both from typical substances such as *salt, water* etc. and from expressions that refer to seas, lakes, oceans and the like. Nouns that qualify as substances are typically inert (or, at least, not lexically specified as moving) and nouns for seas, lakes, oceans etc. denote geographical spaces containing water that may be moving but does not flow in a directional sense. Substances are inherently unbounded ([−b]), expressions referring to seas, lakes and oceans are bounded ([+b]) because they are genuinely delimited by their banks, cliffs or beaches. Similarly, substances can be quantified, as seen in (33a), where the quantification by *more* does not affect the unboundedness of *water*. As for the substance contained in the geographical space identifying a sea, lake or ocean, we note that it may be quantified in a way that remains inaccessible for nouns referring to rivers: the utterance in (33b) is true to fact and means that the overall volume of the Aral Sea has massively shrunk; the utterance in (33c) on the other hand, is not an assessment of the overall volume of the river (which cannot be assessed), but rather of the water level in a particular place (say, the section of the river near the resort where we use to spend our summer holidays).

(33) English
    a. *If you don't want your hydrangeas to dry out, you should give them more water.*
    b. *The North Aral Sea is 40 times smaller in volume than what was once the Aral Sea.*
    c. *There is only half as much water in the river as last year.*

One might point out that the difference between (33b) and (33c) does not truly concern the linguistic expressions, but just the referents they denote. However, while it is true that the difference is grounded in extralinguistic phenomena, it would be wrong to assume that it is not equally relevant to the linguistic expressions themselves. Both *sea* and *river* may be used in a figurative sense: but whereas *sea* may easily be combined with quantified complements (34a) this is not the case for *river* (34b).

(34) English
    a. *a sea of one hundred umbrellas*
    b. ??*a river of one hundred cars*

We take the different behaviour of river names and names of lakes in English with respect to the use of the definite article as an additional indication of their different characterization with respect to lexical features.[16] The lack of a definite article with English names of lakes, as shown in (35), is interesting in two respects. On the one hand, English lake names contrast with names of seas and oceans that take the article. This contrast shows that it is not the mere presence of a common noun in the composition of the proper name that forces the presence of a definite article or not (*Lake Michigan* vs. *the Atlantic Ocean*).[17]

(35)   English
Lake Michigan is one of the greatest lakes in North America.

On the other hand, and more importantly, the striking contrast between lake and river names – especially in English where the presence of the definite article can only be related to the definiteness feature – shows that there must be a fundamental difference in the way lakes and rivers are conceptualized in language. It is our assumption that there is a strong connection between (i) the conceptualization of flowing water, which is relevant for rivers and not for lakes, (ii) the setting of the boundedness feature, with rivers being [αb] and lakes [+b], and (iii) the presence vs. absence of the definite article.

Jackendoff's (1991) nominal typology in (31) primarily operates at the level of the lexicon. Expressions like *cattle* are inherently, i.e. lexically, endowed with the feature specifications [–b], [+i]. Syntactic structures may alter or override the lexical specification. This is ultimately what distinguishes individuals like *a pig* from bare plural aggregates like *buses* (or *pigs*). One syntactic device that is known to alter the lexical specification is the embedding of nouns within DP, and ultimately the interaction of determiners with nouns (among others: Jackendoff 1996; Soh and Kuo 2005). This is exactly what happens in (36), where the lexically unbounded substance *wine* (36a) is converted into a bounded individual through the use of the definite article (36b).

---

[16] An anonymous reviewer points out that English is exceptional in its use of lake names without definite articles. Indeed, the other languages that we have examined thus far all take the definite article with lake names. Whatever the reason for this exceptional behavior of English, it proves once more how truly exceptional the uniform behavior of river names really is.

[17] As pointed out by an anonymous reviewer, the absence of the definite article seems to interact with the position of the common noun: *Lake Michigan* vs. *the Michigan Lake*. Although the observation is correct, there is no easy explanation, since with river names the relative order of common noun and onymic part does not yield a difference with respect to the use of the determiner: *the river Thames* or *the Thames river*.

(36) English
　a. *Mary didn't drink wine.*
　b. *Mary didn't drink the wine.*

If we are correct in assuming that expressions referring to rivers are underspecified for boundedness, hence are lexically specified as [αb], [–i], then the underspecification must be resolved in other, i.e. syntactic, ways in order for such expressions to be adequately interpreted. For common nouns referring to rivers – as for common nouns altogether – this is achieved through the interaction with D. Proper names referring to rivers, i.e. river names, do not move to D and therefore must combine with a determiner that can resolve their underspecification. As we have discussed in Section 4, the definite article is well equipped to fulfill this function because it singles out a unique referent that is known (and prominent) to both speaker and hearer. This often expedites a discourse-prominent interpretation as the ones that were discussed in Section 5.

Consider once again example (21b), repeated as (37): the river name *Seine* is lexically defined as [αb], [–i]. The underspecification for boundedness is uninterpretable and must be resolved syntactically. Since the definite article or, to be more precise, its definiteness feature can convert unbounded into bounded expressions (see (36)), we may safely infer that it converts [αb] into [+b] in (37). The presence of an active definiteness feature now induces a unique referent that is known (and prominent) to both speaker and hearer and therefore may enforce the interpretation whereby the referent of *Seine* is limited to the discourse-prominent section of the river.

(37) English
　Perhaps it is not such a good idea to visit Paris, while the Seine bursts its banks.

7 Conclusion

In this article we tried to come to grips with the fact that in all Germanic and Romance languages that have prenominal determiners river names are used with a definite article. This behaviour is unparalleled among proper names and shows that river names genuinely constitute a particular class of place names.

We have shown that the realization of the definite article may be enforced by different features, definiteness just being one, and that the feature specification of the definite article diverges between languages. The English definite determiner only carries the definiteness feature and therefore we have argued that the obligatory realization of the definite article with river names can only be related to this feature and its function.

The syntactic property of river names to always occur with a definite article seems to match up with a specific semantic property as well. River names, though being proper names, do not qualify as rigid expressions and their interpretation may strongly depend on and vary according to contextual factors. This holds both for proper and common nouns referring to rivers. We have claimed that the particular semantic status of these expressions is ultimately connected with the fact that they are not inherently, i.e. lexically, defined for boundedness. We concluded that the definite article constitutes a syntactic means to resolve the underspecification of the boundedness feature of river names and as such can induce discourse related interpretations.

Acknowledgements

We are grateful to Hartmut Haberland and Lars Heltoft for providing us with information about river names in Danish. We also thank the audience of the *Proper names versus common nouns* workshop at the 41th DGfS conference in Bremen, the three anonymous reviewers for their valuable and much appreciated comments and the editors of this volume. Needless to say that any remaining error is ours.

Abbreviations

1, 3	first, third person
DP	determiner phrase
Du.	Dutch
En.	English
Fr.	French
Fs.	Frisian
Ge.	German
GEN	genitive
It.	Italian
NP	noun phrase

PRS present tense
Pt. Portuguese
SG singular
Sp. Spanish

References

Abney, Steven Paul. 1987. The English noun phrase in its sentential aspect. Cambridge, MA: MIT PhD thesis.
Ackermann, Tanja. 2018. *Grammatik der Namen im Wandel: Diachrone Morphosyntax der Personennamen im Deutschen* (Studia Linguistica Germanica 134). Berlin & Boston: de Gruyter.
Baerman, Matthew, Dunstan Brown & Greville G. Corbett. 2017. *Morphological complexity* (Cambridge Studies in Linguistics 153). Cambridge: Cambridge University Press.
Baker, Mark C. 2003. *Lexical categories. Verbs, nouns, and adjectives* (Cambridge Studies in Linguistics 102). Cambridge & New York: Cambridge University Press.
Berezowski, Leszek. 1997. Iconic motivation for the definite article in English geographical proper names. *Studia Anglica Posnaniensia* 32. 127–144.
Brame, Michael. 1982. The head-selector theory of lexical specifications and the nonexistence of coarse categories. *Linguistic Analysis* 10(4). 321–325.
Bruening, Benjamin. 2020. The head of the nominal is N, not D: N-to-D Movement, hybrid agreement, and conventionalized expressions. *Glossa: A Journal of General Linguistics* 5(1). 15.
Carlson, Greg N. 1977. A unified analysis of the English bare plural. *Linguistics and Philosophy* 1(3). 413–456.
Carlson, Greg N. 1980. *Reference to kinds in English*. New York: Garland Publishing.
Caro Reina, Javier. 2014. The grammaticalization of the terms of address *en* and *na* as onymic markers in Catalan. In Friedhelm Debus, Rita Heuser & Damaris Nübling (eds.), *Linguistik der Familiennamen* (Germanistische Linguistik 225–227), 175–204. Hildesheim: Olms.
Caro Reina, Javier. 2020. The definite article with place names in Romance languages. In Nataliya Levkovych & Julia Nintemann (eds.), *Aspects of the grammar of names: Empirical case studies and theoretical topics* (LINCOM Studies in Language Typology 33), 25–51. München: LINCOM.
Caro Reina, Javier. this volume. The definite article with personal names in Romance languages.
Chierchia, Gennaro. 1998. Reference to kinds across languages. *Natural Language Semantics* 6(4). 339–405.
Chierchia, Gennaro. 2003. Partitives, reference to kinds and semantic variation. In Javier Gutiérrez-Rexach (ed.), *Semantics. Critical concepts in linguistics. Volume 3: Noun phrase classes*, 415–446. London: Routledge.
Clements, George Nick. 1985. The geometry of phonological features. *Phonology Yearbook* 2. 225–252.
Corbett, Greville G. 1991. *Gender*. Cambridge: Cambridge University Press.

Corbett, Greville G. (ed.). 2015. *The expression of gender* (The Expression of Cognitive Categories 6). Berlin & New York: Mouton de Gruyter.
Diesing, Molly. 1992. *Indefinites* (Linguistic Inquiry Monographs 20). Cambridge, MA: MIT Press.
Dobrovie-Sorin, Carmen & Ion Giurgea. 2015. Weak reference and property denotation. Two types of pseudo-incorporated bare nominals. In Olga Borik & Berit Gehrke (eds.), *The syntax and semantics of pseudo-incorporation* (Syntax and Semantics 40), 88–125. Leiden & Boston: Brill.
D'hulst, Yves. 2007. Nominal predicates. In Gabriela Alboiu, Andrei A. Avram, Larisa Avram & Daniela Isac (eds.), *Pitar Moş. A building with a view. Papers in honour of Alexandra Cornilescu*, 87–98. Bucharest: Editura Universităţii din Bucureşti.
Dudenredaktion. 2010. *Duden*. Volume 10: *Das Bedeutungswörterbuch*. 4th edn. Mannheim & Zürich: Dudenverlag.
Enç, Mürvet. 1991. The semantics of specificity. *Linguistic Inquiry* 22(1). 1–25.
Fuchs, Zuzanna, Maria Polinsky & Gregory Scontras. 2015. The differential representation of number and gender in Spanish. *The Linguistic Review* 32(4). 703–737.
Fukui, Naoki & Margaret Speas. 1986. Specifiers and projection. *MIT Working Papers in Linguistics* 8. 128–172.
Giorgi, Alessandra & Giuseppe Longobardi. 1991. *The syntax of noun phrases. Configuration, parameters and empty categories* (Cambridge Studies in Linguistics 57). Cambridge: Cambridge University Press.
Giusti, Giuliana. 1991. The categorial status of quantified nominals. *Linguistische Berichte* 136. 438–452.
Harley, Heidi & Elizabeth Ritter. 2002a. Person and number in pronouns: A feature-geometric analysis. *Language* 78(3). 482–526.
Harley, Heidi & Elizabeth Ritter. 2002b. Structuring the bundle: a universal morphosyntactic feature geometry. In Horst J. Simon & Heike Wiese (eds.), *Pronouns – Grammar and Representation* (Linguistik Aktuell/Linguistics Today 52), 23–39. Amsterdam & Philadelphia: John Benjamins.
Harris, James W. 1991. The exponence of gender in Spanish. *Linguistic Inquiry* 22(1). 27–62.
Harweg, Roland. 1983. Genuine Gattungseigennamen. In Manfred Faust, Roland Harweg, Werner Lehfeldt & Götz Wienold (eds.), *Allgemeine Sprachwissenschaft, Sprachtypologie und Textlinguistik. Festschrift für Peter Hartmann*, 157–171. Tübingen: Narr.
Heim, Irene. 1982. The semantics of definite and indefinite noun phrases. Amherst, MA: University of Massachusetts Amherst PhD thesis.
Higginbotham, James. 1985. On semantics. *Linguistic Inquiry* 16(4). 547–593.
Higginbotham, James. 1987. Indefinites and predication. In Eric Reuland & Alice ter Meulen (eds.), *The representation of (in)definiteness* (Current Studies in Linguistics 14), 43–70. Cambridge, MA: MIT Press.
Jackendoff, Ray S. 1991. Parts and boundaries. In Beth Levin & Steven Pinker (eds.), *Lexical and conceptual semantics*, 9–45. Oxford: Blackwell.
Jackendoff, Ray S. 1996. The architecture of the linguistic-spatial interface. In Paul Bloom, Mary A. Peterson, Lynn Nadel & Merrill F. Garrett (eds.), *Language and space*, 1–29. Cambridge, MA: MIT Press.
Kripke, Saul. 1980. *Naming and necessity*. Oxford: Oxford University Press.

LaPorte, Joseph. 2016. Rigid designators. In Edward N. Zalta (ed.), *The Stanford encyclopedia of philosophy*. https://plato.stanford.edu/archives/spr2018/entries/rigid-designators/ (checked 20/03/09).
Löbner, Sebastian. 1985. Definites. *Journal of semantics* 4(4). 279–326.
Longobardi, Giuseppe. 1994. Reference and proper names: a theory of N-movement in syntax and logical form. *Linguistic Inquiry* 25(4). 609–665.
Longobardi, Giuseppe. 2001. Formal syntax, diachronic minimalism and etymology: The history of French *chez*. *Linguistic Inquiry* 32(2). 275–302.
Longobardi, Giuseppe. 2003. Determinerless nouns: a parametric mapping theory. In Martine Coene & Yves D'hulst (eds.), *From NP to DP*. Volume 1: *The syntax and semantics of noun phrases* (Linguistik Aktuell/Linguistics Today 55), 239–254. Amsterdam & Philadelphia: John Benjamins.
Longobardi, Giuseppe. 2005. Toward a unified grammar of reference. *Zeitschrift für Sprachwissenschaft* 24(1). 5–44.
Longobardi, Giuseppe. 2008. Reference to individuals, persons, and the variety of mapping parameters. In Henrik Høeg Müller & Alex Klinge (eds.), *Essays on nominal determination: From morphology to discourse management* (Studies in Language Companion Series 99), 189–211. Amsterdam & Philadelphia: John Benjamins.
McNally, Louise. 1995. Bare plurals in Spanish are interpreted as properties. In Glyn V. Morrill & Richard T. Oehrle (eds.), *Formal grammar. Proceedings of the Conference of the European Summer School in Logic, Language, and Information, Barcelona, 1995*, 197–212. Barcelona: Polytechnic University of Catalonia.
McNally, Louise. 2004. Bare plurals in Spanish are interpreted as properties. *Catalan Journal of Linguistics* 3. 115–133.
Nübling, Damaris. 2014. *Die Kaiser Wilhelm – der Peterle – das Merkel*. Genus als Endstadium einer Grammatikalisierung – und als Quelle von Re- und Degrammatikalisierungen. In Akademie der Wissenschaften und der Literatur, Mainz (ed.), *Jahrbuch 64 (2013)*, 127–146. Stuttgart: Steiner.
Nübling, Damaris. 2020. *Die Capital – der Astra – das Adler*. The emergence of a classifier system for proper names in German. In Renata Szczepaniak & Johanna Flick (eds.), *Walking on the grammaticalization path of the definite article in German. Functional main and side roads* (Studies in Language Variation 23), 227–249. Amsterdam & Philadelphia: John Benjamins.
Nübling, Damaris. this volume. *Von Heidel- nach Bamberg, von Eng- nach Irland?* 'From Heidel- to Bamberg, from Eng- to Ireland?' On the delimitation of appellative proper names and genuine proper names.
Nübling, Damaris & Fabian Fahlbusch. 2014. *Der Schauinsland – die Mobiliar – das Turm*. Das referentielle Genus bei Eigennamen und seine Genese. *Beiträge zur Namenforschung* 49(3). 245–288.
Nübling, Damaris & Fabian Fahlbusch. 2016. Genus unter Kontrolle: Referentielles Genus bei Eigennamen – am Beispiel der Autonamen. In Andreas Bittner & Constanze Spieß (eds.), *Formen und Funktionen. Morphosemantik und grammatische Konstruktion* (Lingua Historica Germanica 12), 103–125. Berlin & Boston: de Gruyter.
Nübling, Damaris, Fabian Fahlbusch & Rita Heuser. 2015. *Namen. Eine Einführung in die Onomastik*. 2nd edn. Tübingen: Narr.
Pearsall, Judy (ed.). 2002. *Concise Oxford English dictionary*. 10th edn. Oxford: Oxford University Press.

Renzi, Lorenzo. 1976. Grammatica e storia dell'articolo italiano. *Studi di grammatica italiana* 5. 5–42.
Renzi, Lorenzo. 1979. Per la storia dell'articolo romanzo. In Alberto Varvaro (ed.), *XIV Congresso Internazionale di Linguistica e Filologia Romanza (Napoli, 15–20 Aprile 1974)*, vol. 3, 251–265. Amsterdam & Philadelphia: John Benjamins.
Russell, Bertrand. 1905. On denoting. *Mind* 14(56). 479–493.
Saeed, John I. 2009. *Semantics*. Chichester: Wiley-Blackwell.
Salzmann, Martin. 2020. The NP vs. DP debate. Why previous arguments are inconclusive and what a good argument could look like. Evidence from agreement with hybrid nouns. *Glossa: A Journal of General Linguistics* 5(1). 83.
Schmuck, Mirjam. (2020a): The grammaticalisation of definite articles in German, Dutch, and English. A micro-typological approach. In Gunther Vogelaer, Dietha Koster & Torsten Leuschner (eds.), *German and Dutch in contrast: Synchronic, diachronic and psycholinguistic perspectives* (Konvergenz und Divergenz 11), 145–178. Berlin & Boston: de Gruyter.
Schmuck, Mirjam. (2020b): The rise of the onymic article in Early New High German: Areal factors and the triggering effect of bynames. In Renata Szczepaniak & Johanna Flick (eds.), *Walking on the grammaticalization path of the definite article in German. Functional main and side roads* (Studies in Language Variation 23), 199–226. Amsterdam & Philadelphia: John Benjamins.
Soh, Hooi Ling & Jenny Yi-Chun Kuo. 2005. Perfective aspect and accomplishment situations in Mandarin Chinese. In Henk J. Verkuyl, Henriette de Swart & Angeliek van Hout (eds.), *Perspectives on aspect* (Studies in Theoretical Psycholinguistcs 32), 199–216. Dordrecht: Springer.
Szabolcsi, Anna. 1983. The possessor that ran away from home. *The Linguistic Review* 3(1). 89–102.
Thieroff, Rolf. 2000. **Kein Konflikt um Krim*: Zu Genus und Artikelgebrauch von Ländernamen. In Ernest W.B. Hess-Lüttich & H. Walter Schmitz (eds.), *Botschaften verstehen: Kommunikationstheorie und Zeichenpraxis. Festschrift für Helmut Richter*, 271–284. Frankfurt am Main: Peter Lang.
Trubetzkoy, Nikolai S. 1939. *Grundzüge der Phonologie* (Travaux du Cercle linguistique de Prague 7). Prague.
Van Langendonck, Willy. 2007. *Theory and typology of proper names* (Trends in Linguistics 168). Berlin & New York: Mouton de Gruyter.
Vergnaud, Jean-Roger & Maria Luisa Zubizarreta. 1992. The definite determiner and the inalienable constructions in French and in English. *Linguistic Inquiry* 23(4). 595–652.
Vincent, Nigel. 1997. The emergence of the D-system in Romance. In Ans van Kemenade & Nigel Vincent (eds.), *Parameters of morphosyntactic change*, 149–169. Cambridge: Cambridge University Press.
Werth, Alexander. 2014. Die Funktion des Artikels bei Personennamen im norddeutschen Sprachraum. In Friedhelm Debus, Rita Heuser & Damaris Nübling (eds.): *Linguistik der Familiennamen* (Germanistische Linguistik 225–227), 139–174. Hildesheim: Olms.

Johannes Helmbrecht
Proper names with and without definite articles: preliminary results

Abstract: Preliminary results of a typological survey are presented that examines the question of the possibilities to combine proper names with definite articles in a convenience sample of languages with definite articles. The combination of proper names with definite articles may be obligatory, or forbidden, or optional with certain pragmatic effects. Different classes of proper names may be distinguished, but only the most important classes – personal names and place names – are considered in this study, because they presumably are universal classes of expressions. The distribution of definite articles and proper names in the languages of the sample are described, and functional explanations for these patterns are sought.

Keywords: mono-reference, proper names, definite article, classification, emotive connotation.

1 Introduction

The topic of this paper is the variation one finds in noun phrases that contain a proper name as head. These noun phrases require sometimes a definite article in German and English, and sometimes they don't. Compare the examples from English and Standard German for illustration in Table 1.

Table 1: Proper names classes with/without definite articles in English and German.

Type of proper name	English examples	German examples
First Name	Peter	Paul
Nicknames/Hypocoristics	Suzy	Steffi
Last Name	Miller	Müller
Settlement	Cologne	Köln

Johannes Helmbrecht: Universität Regensburg, Fakultät für Sprach-, Literatur- und Kulturwissenschaften, Universitätsstraße 31, D-93051 Regensburg, johannes.helmbrecht@ur.de

https://doi.org/10.1515/9783110672626-005

Type of proper name	English examples	German examples
Country/State	Turkey	Mexico, but **die** Türkei
Region	**the** Black Forest	**der** Schwarzwald
River	**the** Thames/The River Thames	**die** Themse
Mountain	**the** Everest/The Mount Everest	**der** Mount Everest
	the Alps	**die** Alpen
Institution	**the** University of Oxford	**die** Universität Regensburg
	the Louvre	**der** Louvre
Event	**the** French Revolution	**die** Französische Revolution
	the Gulf War	**der** Golfkrieg

Personal names comprise first names (FirstN), last names (LastN) and nicknames (NickN). They do not take a definite article in English and in Standard German. Things are, however, more complicated in Colloquial German (see below). Place names (PlaceN) comprise types of names for settlements (cities, towns, villages and the like), names of countries/states, or regions, names of bodies of water such as rivers, lakes, and so on, names of mountains, and some more. Some of them do not take a definite article, others do. In addition, there are differences between German names and English equivalents: *Turkey* vs. *die Türkei*, for instance, takes a definite article in German, but not in English (also in *Switzerland* vs. *die Schweiz*).

Personal names (anthroponyms) and place names (toponyms) are the most important classes of names (cf. Nübling et al. 2015). They are seen as the prototypical proper names. Names of object, institutions and/or historical events are less prototypical proper names. Trade names and brand names are probably rather common nouns than proper names (cf. Van Langendonck 2007; Nübling et al. 2015).

English and German proper names seem to behave similar with regard to the definite article. Personal names do not take the definite article. This is, however, not the case in other languages. For instance in Armenian and in Greek, personal names require the definite article if they are used in a referential function.[1] Compare the examples in (1) and (2) from Armenian.

---

[1] With regard to personal names, it is important to draw a terminological distinction between referential function and address function. Grammatically, personal names behave differently, if used as address terms. The following examples and almost all examples in this paper illustrate and discuss proper names in referential function.

(1) Armenian (Dum-Tragut 2009: 109)
   Ani-n   **Aram-i-n**   girk'   ē   tal-is
   Ani.NOM-DEF   Aram-DAT-DEF   book.NOM   she.is   give-PTCP.PRES
   'Ani gives a book to Aram.'

(2) Armenian (Dum-Tragut 2009: 109)
   Petros-ĕ   mekn-ec'   **Moskva**
   Petros.NOM-DEF   leave-AOR.3SG   Moscow.NOM
   'Petros left for Moscow.'

The suffixes -n/-ĕ are definite articles. They are allomorphs: -n appears, if the stem ends in a vowel, -ĕ appears, if the stem ends in a consonant. Personal names like *Petros* in Armenian take the definite article. Settlement names such as *Moskva* do not (cf. (2)). In Modern Greek, personal names take obligatorily a definite article, but also all toponyms, even settlement names, compare the examples in (3) and (4).

(3) Modern Greek (Stavros Skopeteas, p.c.)
   o   *'stavros*   *tha*   *mi'lisi s=**to***
   DEF:NOM.SG.M   Stavros:NOM.SG.M   FUT   speak:3.SG LOC=**DEF:ACC.SG.N**
   'berklei
   Berkeley
   'Stavros will speak in Berkeley.'

(4) Modern Greek (Stavros Skopeteas, p.c.)
   o   *'stavros*   *o*   *skope'teas*
   DEF:NOM.SG.M   Stavros:NOM.SG.M   **DEF:NOM.SG.M** Skopeteas:NOM.SG.M
   tha   *mi'lisi*   *s=**to***   *'berklei.*
   FUT   speak:3SG   LOC=**DEF:ACC.SG.N** Berkeley
   'Stavros Skopeteas will speak in Berkeley.'

In (3), the FirstN *stavros* is marked by the definite article *o*, which indicates at the same time case (nominative), number (singular), and gender (masculine). The same holds for the definite article =*to* (accusative, singular, neuter), which is an enclitic to the locative marker *s*. If personal names consist of a FirstN and a LastN, the definite article is used only once having scope over the entire construction. If the definite article is used twice to mark both names individually, this is interpreted as focus marking, cf. (4), symbolized with capital letters.

So, there is some variation within a language and across languages regarding which types of proper names take a definite article, and which ones don't. That proper names take definite articles at all is an astonishing fact vis-à-vis the functional (semantic/pragmatic) properties of proper names and definite articles.

2 Semantic and pragmatic properties of proper names and definite articles

2.1 Proper names

Proper names are a class of referential expressions that are functionally defined. They have the following essential semantic/pragmatic properties. Proper names are mono-referent. They refer ideally to a single unique person/object in the world and are thus inherently singular. See the examples above in Table 1. Even names that are formally marked for plural have a unique singular reference; they refer to places as collectives (*pluralia tantum*); compare the examples in (5) to (8).

(5) English 'the Pyrenees', German 'die Pyrenäen'
(6) **The Pyrenees are** older than **the Alps: their** sediments were first deposited in coastal basins during the Paleozoic and Mesozoic eras. (Source: Wikipedia)[2]
(7) English 'the Philippines', German 'die Philippinen'
(8) ... **the Philippines shares** maritime borders with Taiwan to the north, Vietnam to the west, Palau to the east and Malaysia and Indonesia to the south. (Source: Wikipedia)[3]

The Pyrenees and The Philippines and their German equivalents are PlaceNs that are formally plural. But they refer to a state and to a mountain range as unique entities. That *pluralia tantum* are treated sometimes as singular is shown in the two text examples from Wikipedia. In (6), The Pyrenees are formally treated as plural, see the plural agreement with the auxiliary. However, The Philippines are treated as singular, which can be seen from the 3rd person singular inflection of the verb *share*. In German, The Philippines remain formally plural.

---
[2] Pyrenees. (2017, October 15). In *Wikipedia, The Free Encyclopedia*. Retrieved 14:36, November 6, 2017, from https://en.wikipedia.org/w/index.php?title=Pyrenees&oldid=805443469.
[3] Philippines. (2017, November 5). In *Wikipedia, The Free Encyclopedia*. Retrieved 14:40, November 6, 2017, from https://en.wikipedia.org/w/index.php?title=Philippines&oldid=808862664.

Proper names have a direct reference; they are also called "rigid designators" in the literature (cf. Kripke 1980). Direct reference means that there is a direct relation between the linguistic expression and its referent. There is no concept, notion, or definite description that mediates the reference. Hence, one can say: proper names have no semantic or descriptive content, like definite descriptions; see (9) and (10).

(9) Definite description: *the blue car* (vis-à-vis the cars of a different color)
(10) Proper name: *Paul*

The NP in (9) has descriptive or semantic content in the following sense: the noun *car* designates a category with a certain intension (semantic features) and a certain extension (all entities that fall under this notion). The adjective, on the other hand, restricts the potential class of referents, i.e. cars, to the one that just has this color and the definite article marks it as identifiable by the addressee for various reasons. Nothing of this holds for the FirstN in (10). *Paul* does not designate a class or category. The only semantic content of this PersN is that the name bearer is male.

The relation between a proper name and its referent is conventionalized. Either there is a conventional ceremony, in which a proper name is explicitly accorded to the referent, or the speech community develops a proper name for a referent that is sufficiently important to become named individually. FirstNs of newborns receive their name by a baptism ceremony, but PlaceN are usually not baptized. Because of the conventionalized relation between a proper name and its referent, proper names are inherently definite and specific. Proper names can be used successfully in discourse only, if both, speaker and addressee, are familiar with the referent of the proper name. If the name of a referent is not known, it has to be introduced explicitly, before it can be used as a referential expression. Compare the examples in (11) and (12).

(11) *Paul came over yesterday.*
(12) *We have a new colleague, his name is Paul ...*

The utterance in (11) makes sense only, if the addressee knows the referent of *Paul*. Otherwise, the speaker has to introduce the intended referent and has to provide a name for him/her, as illustrated in (12). There is a broad consensus that proper names are inherently definite (cf. Hawkins 1978; Löbner 1985: 299; Wotjak 1985: 7; Lyons 1999; Van Langendonck 2007: 154; etc.). Morphosyntactic evidence for this view is summarized and presented in Van Langendonck (2007: 154–157).

3 Definite articles

Definiteness is not a language-specific grammatical category, as is presupposed often in descriptive grammars and even studies in general linguistics. It is rather a universal functional domain comprising various semantic and pragmatic notions (cf. König 2018), which in turn could be considered as comparative concepts (in the sense of Haspelmath 2010; see also Croft 2016) for language comparison. The semantic and pragmatic concepts of definiteness are the familiarity and identifiability of the referent, since the intended referent is present in the universe of discourse of the interlocutors. Further concepts that belong to this functional domain and that are discussed in the literature are uniqueness, salience, existence, and inclusiveness (cf. König 2018).

There are typical contexts, in which definite articles (in English and other European languages) appear with the noun (see Hawkins 1978; summarized in Himmelmann 1997: 35–42). Definite articles are used, if the referent is a visible or non-visible part of the situation, in which the utterance takes place. This type of article use is called "immediate situation uses" by Hawkins (1978: 108); cf. the following examples (Hawkins 1978: 111–112).

(13) *Pass me **the** bucket, please!*
(14) *Beware of **the** dog!*

The referents of *the bucket* in (13) and *the dog* in (14) are identifiable for the addressee, because these referents are part of the speech act situation and perceivable.

Furthermore, definite articles are used, if the referent has been mentioned in the preceding discourse. This is called "anaphoric usage"; cf. the example in (15).

(15) *Fred was discussing **an interesting book** in his class. I went to discuss **the book** with him afterwards.* (Hawkins 1978: 86)

The definite article marks the second mention of the *book* in (15) and indicates that the referent is identifiable, because it has been mentioned shortly before in discourse.

A third context for definite articles is given, if the referent belongs to the encyclopedic/pragmatic knowledge of both speech act participants; this is called "larger situation uses" in Hawkins (1978: 115); cf. the examples in (16).

(16) **the** sun, **the** Queen, **the** Prime Minister, etc.

The definite article in these cases is used, because the referents are unique and because it can be presupposed that they are known by the speech act participants and the members of the speech community in general.

A fourth typical context for definite articles is given, if there is a specific culturally mediated or associative relation between a previously mentioned entity and the referent; often this is a part-whole relation. This context is called "associative anaphoric use" in Hawkins (1978: 123); cf. the following example.

(17) Paul bought **a new car**, but he had to replace **the** tires.

The referent of the expression *the tires* is identifiable, because they are part of the entity *a new car* that has been already mentioned in the preceding clause. In addition, there are grammatical contexts, mostly complex noun phrases, that require the definite article obligatorily, compare the examples in (18).

(18) a) a nominal with a relative clause: **the** car Peter has bought
b) a nominal with a complement clause: **the** conviction that he has to vote
c) a nominal with a genitive attribute: **the** tires of the car
d) and some others;

According to Löbner (1985), the different usage contexts mentioned above define different types of definiteness. He distinguishes in this respect semantic and pragmatic definiteness. The question remains: if proper names are inherently definite (and specific), why do they co-occur so often with definite articles. In principle, there would be no functional need to mark proper names as definite. In order to obtain an answer to this question, it is necessary to have a typological overview of the variation that exists with regard to these constructions. In what follows, I will present some very preliminary results of a typological survey that investigates the distribution of definite articles with different types of proper names in languages that have definite articles.

4 Methodological remarks

We compiled a sample of about 370 languages, which have in common that they were reported to have a marker for definiteness no matter whether this marker is a word, clitic or affix. The basis for this sample is the WALS chapter on definite articles by Dryer (2013). From this initial list of languages, we selected a few that are well described and documented and that are distributed geographically over different continents and different language families. All in all, there are data of about 40 languages that were included in this survey. Then we set up a typology of proper name types – similar to the list of name types in Table 1 above. For each language, we tried to answer the following questions:
a) Does the definite article in language x of the sample occur with proper names, or not?
b) Which one of the different proper name types takes a definite article, and which ones don't?
c) Are the definite articles in these constructions obligatory, optional, or forbidden?
d) If the definite article is optional, what kind of semantic/pragmatic effects can be identified, if the definite article is attached to the PN?

Some difficulties arose, though. First, explicit structural and distributional information on proper names is often lacking in descriptive grammars. Often, we do not find any information or explicit examples of noun phrases containing proper names in different syntactic contexts (e.g. in different grammatical relations). Often proper names appear in descriptive grammars only in form of lists. Second, if there is some grammatical information on proper names, then this information often is restricted to personal names; there is rarely explicit information on toponyms; and the entire range of names is almost never covered. One reason may be, of course, that certain name types do not occur in the languages. For instance, many small-scale societies do not have last names.[4] The

---

[4] A good amount of data I used for this study stem from questionnaires which were filled out by language experts (either native speakers of a language, or linguistics working on that language). It turned out that some of the experts told us that there are no last names in their language of expertise. Others provided us with examples using last names of the culturally dominant language surrounding the minority language. For instance, Angelika Jakobi and Elsadig Omda who provided us with data from Beria (Saharan language of Darfur, Sudan) told us that there are no last names in Beria. Similar responses came from experts of North and South

other reason is that authors of descriptive grammars do not pay attention to the possible morphosyntactic differences between proper names and common nouns. This lack of information in grammars, or of name categories in languages has been dealt with in two ways in this study. First, we collected all information we could get on the use of definite articles with proper names in a language of the sample. However, for the typological analysis in §4, we lumped together the different categories to two broad name categories, personal names and place names. I did not take in the information on other proper name types we obtained for a few languages into the analysis.

With these restrictions in mind, we can come up with the following preliminary results. The sample of languages and the sources of information on the proper name constructions with and without a definite article are given in the Appendix 0 below.

5 Typological survey

5.1 The definite article with personal names

Personal names may be used to address people or to refer to people. This distinction is reflected in the grammar of personal names, more specifically, in the ability to take the definite article. In general, personal names in address function do not take the definite article.[5] Thus, the following survey of personal names only deals with names in a referring function.

There is a large group of languages in our sample, which do not take the definite article with personal names. This holds for all name categories such as first names, nicknames/hypocoristics, and last names, if they have any.

American indigenous languages. Often, last names are taken from the culturally dominant language, e.g., from Spanish or English.

5 One of the reviewers mentioned that there are exceptions to this rule. Some Austronesian languages take the definite article with personal names in the vocative, and even Portuguese optionally allows the definite article with personal names in address function with particular politeness effects.

Table 2: Personal names without definite article.

#	Language	Genetic affiliation
1.	Tuvaluan	Austronesian, Oceanic
2.	Jamsey	Dogon
3.	Gumuz	Gumuz
4.	English	Indo-European, Germanic
5.	German	Indo-European, Germanic
6.	French	Indo-European, Romance
7.	Spanish	Indo-European, Romance
8.	Persian	Indo-European, Iranian
9.	Chimariko	Isolate of the US
10.	Masalit	Maban
11.	Supyire	Niger-Congo, Gur
12.	Akan	Niger-Congo, Kwa
13.	Noon	Niger-Congo, Northern Atlantic
14.	Lepcha	Sino-Tibetan, Lepcha
15.	Assiniboine	Siouan
16.	Hoocąk	Siouan
17.	Koyra Chiini	Songhay
18.	Hungarian	Uralic
19.	Nuuchahnulth	Wakash

These languages correspond to our expectation that proper names in general, and personal names in particular do not need to be marked by a definite or specific article, because they possess this property already inherently. Note that colloquial varieties of these languages, in particular spoken German, may deviate from this rule (see below).

However, there is also a number of languages that do not correspond to our expectations. These languages – presented in Table 3 – obligatorily require the definite article. They are from different regions and language families.

Table 3: Personal names with definite article.

#	Language	Genetic affiliation
1.	Tukang Besi	Austronesian, Celebic

#	Language	Genetic affiliation
2.	Qaqet	Baining, Papua New Guinea
3.	Armenian	Indo-European
4.	Greek	Indo-European
5.	Seri	Isolate of Mexico[6]
6.	Hidatsa	Siouan
7.	Tepehuan	Uto-Aztecan

Examples from Greek were given above in (3) and (4). Compare the examples from Armenian, Hidatsa, Seri, Tepehuan, Tukang Besi, and Qaqet below. In Armenian, see (19) and (20), the definite article is obligatory with a personal name, but can be replaced by an indefinite marker, if the referent of the PersN is not known by the speech act participants.[7]

(19) Armenian (Dum-Tragut 2009: 109)
    Petros-ĕ   *mekn-ec'*   *Moskva*
    Petros-DEF   leave-AOR.3SG   Moscow.NOM
    'Petros left for Moscow.'

(20) Armenian (Dum-Tragut 2009: 109)
    Mi (inč'or)   **Petros**   *mekn-ec'*   *Moskva*
    A (certain)   **Petros**   leave-AOR.3SG   Moscow.NOM
    'A (certain) Petros left for Moscow.'

In general, Hidatsa personal names take the definite article, but this holds only for native Hidatsa names. Personal names borrowed from English, do not take the definite article. See the examples in (21) and (22). Thus, foreignness plays a role for the distribution of definite article with personal names (see below for further grammatical and pragmatic factors).

---

[6] In Marlett (2008:49), it is reported that Otomi languages (Otomanguean) and Nahuatl (Uto-Aztecan) require the definite article with personal names, too. I could not check this claim yet. However, one of the reviewers indicated, that Classical Nahuatl, Colonial Nahuatl, and Modern Nahuatl personal names usually are accompanied by the specifier *in* (or variants of it), whereas place names never take this specifier. This rule does not hold in modern Huasteca Nahuatl, though.

[7] One of the reviewers suspected that this is always the case, if a language has indefinite articles that indicate specificity.

(21) Hidatsa (Park 2012: 392)
   Cagáàga-mìà-s   se'-ri   séé-c
   bird-woman-DEF that-ERG   say-DECL
   'Bird Woman said that.'

(22) Hidatsa (Park 2012: 391)
   Grace   sé'-g   mada-magi-mácha-'a-c
   Grace   that-CRD   1POSS-RECIP-sibling-PL-DECL
   'Grace and I are siblings.'

The following examples all illustrate the claims in Table 3 that these languages use the definite article with personal names.

(23) Seri (Marlett 2008: 49)
   Hipíix   **Juan**   **quih**   haa   ha.
   this.one   **Juan**   DEF   one.that.is   DECL
   'This is Juan.'

(24) Southeastern Tepehuan (Willett 1991: 239)
   Ticca-'-ap   **gu**   **Juan**   na   pai'dyuc   va-r-jimda-m   para   Corian
   ask-FUT-2SG   **def** **John**   SUB   when   RLZ-EXS-go-DES   to   Durango
   'Please ask John, when he wants to leave for Durango City.'

(25) Tukang Besi (Donohue 1999: 305)
   [Te   [[La Ode Wuna]_{N'}]_{NP}]_{RP}   no-rato   kua   Buru.
   CORE   La.Ode.Wuna   3R-arrive   ALL   Buru
   'Lord Wuna arrived at Buru.' (W: 17)

(26) Qaqet (Hellwig, p.c.; Hellwig 2019: 81–83)
   deqialu   ma**sirini**   mramagulengga
   dekia=lu   **ma**=**sirini** met=**ama**=guleng-ka
   CONJ   3SG.F.SBJ=see   ART.ID=<NAME>   IN= ART=malay_apple-NC.SG.M
   'and she saw **Sirini** in the malay apple tree' (N11AESSirini-0008)

In addition, there are many languages, in which we find a kind of mixed situation in the sense that certain grammatical or pragmatic factors constrain or determine the use of the definite article with personal names. Either the definite article is required with personal names in different regions, dialects, or varieties of a standard language, or the use of the definite article depends on the proper name type, or the syntactic construction. The different situations we find are summarized in Table 4 and will be illustrated briefly in turn.

Table 4: Definite articles with personal names: pragmatic and grammatical factors.

#	Factor (General)	Specific	Language
1.	variety	colloquial vs. standard variety	Coll. German, Coll. Italian, Coll. Hungarian
2.	pragmatic	social/emotional closeness; negative connotation;	Coll. German, Coll. Italian
3.	foreignness	native PerN vs. borrowed PersN	Hidatsa
4.	personal name type	NickN vs. FirstN/LastN	Bulgarian, Arabic
		FirstN vs. LastN	Kabardian
		baptismal name	Portuguese
5.	sex/gender of name bearer	female vs. male PersN	Italian, Savosavo
6.	name construction/complex names	FirstN LastN-DEF-ABL	Dime
		FirstN DEF LastN	Breton
7.	syntactic construction	modification	German, English, Italian

Line 1 and 2 of Table 4. In Standard German, no definite article is used with personal names; neither with the FirstN, nor with the LastN (see Table 2). However, in spoken (non-standard) German, there is a significant difference between Southern (and Middle) Germany and Northern Germany. In the South, the usage of FirstN with the definite article is almost the rule. No emotive or expressive meaning is associated with it. In the North, the use of the definite article with FirstN/LastN is associated with negative attitudes of the speaker towards the referent (cf. Bellmann 1990: 257–293; Nübling et al. 2015: 126; Berchtold and Dammel 2014; Werth 2014; and recently Werth 2020). Similarly, in Italian: Standard Italian does not allow the definite article with personal names, but there are varieties in Northern Italy that require the article; cf. (27). See also example (34).

(27) Italian (Ilenia Tonetti, p.c.)
L(a)' Ilenia non è ven-ut-a a scuola oggi.
DEF.F I. not AUX.PST come-PTCP-F to school today
'**The** Ilenia did not come to school today.'

The presence of the definite article in (28) before the personal names signals social or emotional proximity/familiarity. This usage occurs only in Northern varieties of Italian (Ilenia Tonetti, p.c.).

(28) Italian (Maiden and Robustelli 2000: 69)
    A   teatro   ci   anda-va-no   di_solito mia   madre,
    to   theatre there   go-PST-3PL   usually   my   mother
    la   Paola   e   Mario.
    la   Paola   e   Mario.
    DEF.F.SG P.   and M.
    'My mother, Paola and Mario usually went to the theatre.'

Line 3 of Table 4: only native personal names are used with the definite article in Hidatsa (Siouan), borrowed personal names from English are used without the definite article thus copying the rules in English.

Line 4 of Table 4. In Bulgarian, FirstN and LastN are not allowed with the definite article, but with NickNs, the definite article is obligatory; cf. (29) and (30).

(29) Bulgarian (Petar Kehayov, p.c.)
    Marija   mi   e   priatelka.
    Maria   1SG.DAT is   friend
    'Maria is my friend.'

    *Marija-ta mi   e   priatelka.
    Maria-DEF.F 1SG.DAT is   friend

(30) Bulgarian (Petar Kehayov, p.c.)
    Tesla-ta   pristigna   včera.
    adze-DEF.F arrive:PST.3SG yesterday
    'Adze ("hammering tool" as nickname) arrived yesterday.'

In Kabardian, FirstNs are not case marked, at least not by the core cases NOM (-*r*) and ERG (-*m*). However, LastNs receive the case endings. Matasović explains this by the fact that NOM and ERG case ending also encode definiteness (cf. Matasović 2010: 19–22). Unfortunately, there are no illustrating examples of LastN in Matasović grammar. The FirstNs in both examples (32) and (33) do not receive the NOM -*r*, which would appear with common nouns (see (31)). They are exempted from case marking at all.

(31) Kabardian (Matasović 2010: 19)
    sa   **txəɬa-r**   *q'a-s-śt-ā-ś*
    1SG.NOM/ERG   book-NOM   DIR-1SG-take-PST-AFFIRMATIVE
    'I took the book.'

(32) Kabardian (Matasović 2010: 22–23)
    Maryan   *səma*   *mā-kʷa*
    Maryan   ASSOC.PL   3PL-go
    'Maryan and the others are going.'

(33) Kabardian (Matasović 2010:22f)
    Maryan   *səma*   *s-aw-ɬāġʷ*
    Maryan   ASSOC.PL   1SG-PRES-see
    'I see Maryan and the others.'

One of the reviewers pointed out to me that in Portuguese, baptismal names are used with the definite article for pragmatic reasons (cf. Cunha and Cintra 1984: 225–226).

Line 5 of Table 4. Sometimes, the usage of the definite article is dependent on gender. For instance, in Italian, the definite article is used with LastNs in case that the individual referred to is female. The definite article is usually not used with LastN referring to males; see the example in (34).

(34) Italian (Maiden and Robustelli 2000: 68)
    Dev-o   *parl-are*   *con **la***   **Pisani**.
    must-1SG   speak-INF   to DEF.FEM Pisani
    'I have to speak to the Pisani.'

Savosavo, a Papuan language of the Salomon Islands, has a feminine definite article ("determiner" DET in Wegener's terms) that is specifically used with female personal names, cf. the example in (35) with the female personal name *Poluku*.

(35) Savosavo (Wegener 2012: 85)
    Ko   *nyba*   *ko-va*   *nini=e*   **koi**   **Poluku**
    DET.SG.F child   3SG.F-GEN.M name=EMPH DET.SG.F Poluku
    'The daughter's name was Poluku.'

In addition, the same determiner *koi* DET is used with question words that ask for the name of a woman (see (36) below). This definite article/determiner is not used with male names.

(36) Savosavo (Wegener 2012: 86)
    Koi   *ai=e*   *no=na?*
    DET.SG.F   who=EMPH   2SG=NOM
    'Who are you? (asking a woman)'

Line 6 of Table 4. The use of the definite article with a personal name may also be dependent on the syntactic construction, in which the name is a part of. In Dime, an Omotic language of Ethiopia, FirstN do not co-occur with the definite article; cf. (37). There are two FirstNs, one (*šiftaye*) is marked by the nominative (zero), and the second (*maikro*) by a combination of oblique case markers (comitative-directional-ablative/COM-DIR-ABL).

(37) Dime (Seyoum 2008: 102)
    koos-is-im   *šiftaye*   *maikro-ka-bow-de*   *gis'-i-n*
    ball-DEF-ACC   **shiftaye**   **maikro**-COM-DIR-ABL   kick-PF-3
    'Shiftaye kicked the ball away from Maikro.'

However, if the FirstN is combined with the LastN of the individual, the definite article *(-is/-iz)* is used with the LastN plus the ablative (abl).

(38) Dime (Seyoum 2008: 188)
    ʔaté   *ʔi-sko*   *bábé*   ***šiftaye***   ***mihel-is-de***
    1SG.SUBJ   1S.OBJ-GEN   father   shiftaye   mihel-DEF-ABL
    'My father is Shiftaye Mihel, ...'

A very similar construction is found in modern Breton. The FirstN alone does not receive a definite article, but if FirstN and LastN are combined, the definite article precedes the last name; cf. (39). The same holds, if the LastN combines with a title. In this case, the definite article precedes the entire construction; cf. (40).[8] Note that the definite articles in (39) and (40) are allomorphs.

---

[8] One of the reviewers rightly points out that there are similar tendencies in spoken German. The use of titles such as *Herr* 'Mr.' and *Frau* 'Mrs.' plus LastN facilitate the use of the definite article, e.g., *der Herr Schmidt* 'the Mr. Smith'.

(39) Modern Breton (Press 1986: 76)
    Loeiz   ***ar***   *Gov*
    L.   DEF   G.
    'FirstN DEF LastN'

(40) Modern Breton (Press 1986: 76)
    An   *Aotrou* *Gov* ...
    DEF   Mr.   Gov
    'Mr. Gov ...'

Line 7 of Table 4. It seems that other syntactic constructions also play a role with regard to definiteness marking. In German and English, the definite article is obligatory, if the personal name is modified by an attribute (e.g. adjectival attribute, prepositional attribute); see the German examples in (41) and (42) with the English translations.

(41) **der** schlaue Peter '**the** smart Peter'
(42) **der** Peter mit der Brille '**the** Peter with the glasses'

There are also some languages in our sample, which allow the definite article occasionally with personal names, but no specific communicative effect is reported; cf. Table 5.

Table 5: Personal names appear occasionally with definite article.

#	Language	Genetic affiliation
1.	Goemai	Afro-Asiatic, West Chadic
2.	Ingessana	Eastern Sudanic, Eastern Jebel
3.	Qiang	Sino-Tibetan, Qiangic
4.	Bilua	Solomons East Papuan, Bilua
5.	Lavukaleve	Solomons East Papuan, Lavukaleve
6.	Mian	Trans-New Guinea, Ok

5.2 The definite article with place names and personal names

The last section examined the occurrence and non-occurrence of personal names with definite articles. Three types of languages were found so far:

a) languages that do not allow the definite article with personal names (cf. Table 2)
b) languages that obligatorily require the definite article with personal names (Table 3)
c) and languages, in which the definite article co-occurs with personal names depending on various syntactic, semantic or pragmatic or sociolinguistic conditions (cf. Table 4).

The next question that will be examined is: how do place names in the languages of these three groups behave with regard to the definite article? It should be noted that no distinction was made whether place names appear within an adpositional phrase or not. One of the reviews rightly pointed out that the definite article may be obligatorily omitted within a prepositional phrase (cf. Caro Reina 2020; but also, Stolz and Levkovych this volume).[9] I'll start with type (a) languages, i.e. languages that do not allow the definite article with personal names; cf. Table 6.

Table 6: The definite article and place names in type (a) languages.

#	Personal names with no DEF	Genetic affiliation	Place names –/+DEF
1.	Tuvaluan	Austronesian, Oceanic	no DEF with place names
2.	Jamsey	Dogon	no DEF with place names
3.	Gumuz	Gumuz	no DEF with place names
4.	English	Indo-European, Germanic	DEF with many/most place names except settlement names and country names
5.	German	Indo-European, Germanic	DEF with all place names except settlement names and country names
6.	French (similar in Portuguese, and Spanish)	Indo-European, Romance	DEF with all place names except settlement names; another exception is the construction *en* 'in' + Name of region/country (*en Chine, en Russie*, etc.)
7.	Chimariko	Isolate	no DEF with place names

---

9 See Stolz et al. (2017, 2018) for detailed typological surveys of the morphosyntax of place names.

#	Personal names with no DEF	Genetic affiliation	Place names −/+DEF
8.	Masalit	Maban	no DEF with settlement names (no info for other types)
9.	Supyire	Niger-Congo, Gur	no DEF with place names
10.	Akan	Niger-Congo, Kwa	no DEF with place names
11.	Noon	Niger-Congo, Northern Atlantic	no DEF with place names
12.	Lepcha	Sino-Tibetan, Lepcha	no DEF with place names
13.	Assiniboine	Siouan	no DEF with place names
14.	Hoocąk	Siouan	no DEF with place names
15.	Nuuchahnulth	Wakash	no DEF with place names
16.	Koyra Chiini	Sonhgay	no DEF with place names
17.	Hungarian	Uralic	DEF with all place names except settlement names and country names

A majority of the languages (13 out of 17 type (a) languages) that do not use the definite article with PersNs do not use the definite article with PlaceNs either. But, on the other hand, it is obvious that (some) European languages (English, German, French in our sample; see shaded languages in Table 6) behave quite differently. The definite article is generally used with place names except settlement names and occasionally country/state names. This also holds for Hungarian, which is, of course, not Indo-European. One might hypothesize that this is perhaps an areal feature of European languages. However, more research is necessary to confirm this hypothesis.

The next group to examine are type (b) languages, i.e., the languages that obligatorily require the definite article with personal names; cf. Table 7.

Table 7: The definite article and place names in type (b) languages.

#	Personal names with DEF	Place names +/− DEF
1.	Greek	DEF with all place names
2.	Hidatsa	DEF with all place names
3.	Seri	DEF with all place names
4.	Qaqet	DEF with all place names
5.	Armenian	no DEF with place names (examples in the grammar only for settlement names, and county names)

#	Personal names with DEF	Place names +/− DEF
6.	Tukang Besi	no DEF with place names
7.	Tepehuan	no DEF with place names

With regard to place names, the behaviour of more than half of the languages of type (b) languages in Table 7 is quite consistent: Greek, Hidatsa, Seri, and Qaqet require the definite article with place names as well as with personal names. The interesting cases are Modern (Eastern) Armenian, Tukang Besi, and Tepehuan; all three do not use the definite article with place names, only with personal names; cf. the examples in (43) for Armenian, (44) for Tukang Besi, and (45) for Tepehuan.

(43) Modern Armenian (Dum-Tragut 2009: 86)
 Gařnan-ě   gnal-u   enk'   **Moskva**
 spring.DAT-DEF go-PTCP.FUT we.are Moscow.NOM
 'In spring we will go to Moscow.'

(44) Tukang Besi (Donohue 1999: 305)
 [Te   [[La Ode Wuna]$_{N'}$]$_{NP}$]$_{RP}$ no-rato **kua** **Buru**.
 DEF   La.Ode.Wuna   3R-arrive ALL   Buru
 'Lord Wuna arrived at Buru.' (W: 17)

(45) Tepehuan (Willett 1991: 194)
 Aptuvús ta'm-ach-ich va-jí mi'   **dyɨr** **Vódamtam** **para** **Corian**
 bus   on-1PL-PRF   RLZ-go there from   Mezquital   to   Durango
 'We went by bus from Mezquital to Durango.'

The Armenian case is particularly interesting, because the definite article is used for place names, if they do not refer to places as geographical entities, but to social and legal entities as is the case in example (46). If the place name *Hayastan* 'Armenia' is meant geographically, it is used without definite article, compare (47).

(46) Armenian (Dum-Tragut 2009: 94)
 Rusastan-ě   ew   **Hayastan-ě**   li   en   včřakanut'y-amb
 Russia.NOM-DEF   and Armenia.NOM-DEF   full they.are resolutions-INST
 'Russia and Armenia are full of resolutions(s).'

(47) Armenian (Dum-Tragut 2009: 96)
   Hayastan-ic'   amen   gn-ov   petk'ēheřan-a
   Armenia-ABL   all   price-INST   leave-DEB.FUT.3SG
   'He must leave Armenia at all costs.'

The two place names in (46) *Rusastan* 'Russia' and *Hayastan* 'Armenia' refer rather to the governments of the two countries than to geographical regions. They both are marked by the definite article.

The third type (c) of languages includes the languages, in which the definite article co-occurs with personal names depending on various syntactic, semantic or pragmatic conditions. Compare Table 8 for how these languages behave with regard to place names.

Table 8: Some languages of group (c).

#	Different factors for the use of definite article with personal names	Genetic affiliation	Place names +/−DEF
1.	Coll. German	Indo-European, Germanic	no DEF with settlement names and some country names, but DEF with all other place names
2.	Italian	Indo-European, Romance	no DEF with settlement names,10 but DEF with all other place names
3.	Bulgarian	Indo-European, Slavic	no DEF with settlement names and country/region names; rare with river names, frequent with lakes, impossible with ocean names;
4.	Savosavo	Solomons East Papuan, Savosavo	No DEF with place names (we have examples for settlement and country/region names);

---

10 One of the reviewers presents some Italian settlement names such as *Il Cairo, La Spezia, L'Aquila, L'Aia, La Mecca* that take the definite article. I don't think that these examples are real counter-examples to the rule. In these cases, the definite article is just lexicalized and has become an integral part of the toponym. Similar cases can be found also in other languages, e.g., in *The Hague/Den Haag*.

Spoken German in line 1. in Table 8 is not different from Standard German with regard to the combination of definite article and place names. The definite article is used with all place names except settlement names and some country names. Similarly, in Italian (see line 2.). The definite article is used with all place names except settlement names. However, as soon as there is an attribute modifying the place name, the definite article is obligatory. See (48).

(48) Italian (Maiden and Robustelli 2000: 68–69)
Parigi   ha   oggi   un   po'   gli   stessi   *problem-i*
Paris   has.PRS   today INDEF.M.SG few   DEF.M.PL same   problem-M.PL
de-lla   *Parigi*   *medievale*
PREP-DEF.F.SG Paris   medieval
'Paris today has rather the same problems, **as medieval Paris did.**'

The same is true for Standard and Coll. German but compare the English translation of (48). No definite article is used with the second place name with the modifying adjective. In Bulgarian (see line 3.), the situation is somewhat different. No DEF is used with settlement names, country/region names, and ocean names. The definite article is rare with river names, but frequent with lake names.

(49) Bulgarian (Petar Kehayov, p.c.)
*Utre   *pătuvam*   za   *Sofija-ta*
tomorrow   travel:PRS.1SG   to   Sofia-DEF.F
'Tomorrow, I travel to (*the) Sofia.'

This finding does not allow to draw any conclusion, it could be an idiosyncrasy of Bulgarian. Savosavo (see line 4.) does not allow the definite article with place names.

6 Discussion of the data

There are several quite explicit claims with regard to the co-occurrence of proper names and definite articles that can be deduced from the available literature on proper names. One claim is that personal names are the prototypical names, and all other name types, i.e., animal names, place names, names of institutions, names of events, etc. are less prototypical (see Nübling et al. 2015: 104). Similarly, Van Langendonck claims that personal names and place names are the most prototypical name types (cf. Van Langendonck 2007: 171–173). A simplified version of this claim is presented in (50a).

Another claim applies to place names only. It is hypothesized that there is a sub-hierarchy among place names according to the importance of the named place to humans such that place names with a high degree of involvement and interaction of humans in and with a place are prototypical, and place names with a low degree of involvement of humans are less prototypical (cf. Van Langendonck 2007: 202–212). According to this semantic hierarchy, settlement names are more prototypical than river or mountain names, because settlements are highly salient for human inhabitants and constitute a place of a high degree of interaction. This is not the case with mountains and rivers, for instance. A simplified version of this hierarchy is given in (50b)

(50) a. Hierarchy I: personal names > place names > other names
 b. Hierarchy II: settlement names > country/state names > regions > other[11]

Since both semantic hierarchies relate different name categories to the gradual notion of a prototype category, it may be hypothesized that there are differences in formal markedness that correspond to the differences in prototypicality. More prototypical categories are less marked and less prototypical categories are more marked. If formal markedness differences correspond to the proposed differences in prototypicality, this may count as evidence for the validity of the postulated hierarchies.

The occurrence of proper names with definite articles is just one formal sign of markedness in Van Langendonck's theory (2007).[12] More specifically, Van Langendonck claims that the prototypical name types such as personal names (Hierarchy I) and settlement names (Hierarchy II) are generally zero marked, i.e., do not occur with the definite article (cf. Van Langendonck 2007: 172). I will come to this claim later. With regard to place names (i.e., Hierarchy II), Van Langendonck identified other formal means of markedness; cf. Table 9.

---

11 See Caro Reina (2020) for a critical discussion of this hierarchy.
12 Note that Van Langendonck's prototype theory of proper names is conceptually based on Croft's Radical Constructions Grammar and his conception of typological markedness and prototypicality (cf. Croft 2001).

Table 9: Markedness scale for place names (cf. Van Langendonck 2007: 205–210).

	Less marked		More marked	
Formal scale	zero	suffix	article	classifier (+article)
Semantic scale	settlement	states, countries	regions, rivers, lakes, forests, mountains, etc.	rivers, lakes, oceans, forests, mountains, valleys, etc.
English/German examples	London Köln	German-**y** Tschech-**ien**	**the** Thames **der** Rhein	the Mount Everest der Schwarzwald

The English and German examples in Table 9 illustrate Van Langendonck's (2007: 202–212) idea that the more prototypical place names are formally less marked than the non-prototypical place names. Mountain names, for instance, are marked by a so-called classifier noun like *Mount* and a definite article as in *the Mount Everest*.[13] Likewise, in German with *der Schwarzwald* 'the Black Forest', where the classifier noun is *-wald* 'forest'. Names for states/countries have in common that they designate socially and politically bounded units comprising sets of settlements. Formally, they don't take the definite article (there are exceptions, though), but often do have a suffix that can be interpreted as designating a collective unit, and are thus less marked than river names, mountain names, etc.

Van Langendonck's claim that personal names and place names are the most prototypical name types and are thus the least marked names types is incorporated in the two hierarchies in (50). He explicitly claims that these two name types are basically zero marked, also with respect to the definite article (see Van Langendonck 2007: 172). There is a significant number of languages examined for this survey that confirm this idea. But many don't and perhaps systematic exceptions occur.

All languages of the sample are article languages, i.e., they are described as having definite articles (see the appendix). A great number of languages of the sample, almost 50%, does not use the definite article with personal names, a result which would be expected by the definite nature of proper names and the prototype view on personal names presented in (50a). However, a little bit less than 25% of the languages require the definite article, and there are approximately another 25% of languages that require the definite article for personal names depending on various syntactic, semantic, and pragmatic contexts. The

---

[13] See also Stolz and Levkovych (2020) on this kind of classifiers.

claim that personal names are always zero coded has to be rejected vis-à-vis these data.

The additional claim that personal names are less marked than place names and other less typical name types can be confirmed only to some degree. It is the European languages (the standard varieties of English, German, French, and Hungarian) that show the cline in markedness predicted in (50a). They do not take the definite article with personal names, but with all kinds of place names except settlement names and country names. The other languages that do not take definite articles with personal names do so also with place names, i.e., don't take definite articles with place names either. With regard to the definite article, no markedness differences can be determined.

However, there are a few languages that show markedness patterns that go completely against the predictions in the hierarchy in (50a). In Armenian, Tukang Besi, and Tepehuan we have definite articles with personal names, but not with place names (and probably not with other less prototypical names). These data cannot be explained by means of the prototypicality theory of proper names. In addition, there are some languages that take the definite article with personal names and do so also with place names (i.e., Modern Greek, Seri, Hidatsa, Qaqet). No markedness asymmetry can be determined in these cases either.

With regard to the second markedness hierarchy – the sub-hierarchy of place names – proposed by Van Langendonck, the following results can be obtained. Markedness differences between different types of place names are obvious only with European languages. For other languages outside this area, we find two different marking possibilities. On the one hand, languages that do not take the definite article with personal names, do so also consistently with place names. On the other hand, languages that do require the definite article with personal names also require the definite article for all place names consistently. In fact, both groups of languages either require the definite article for personal and place names, or do not allow the definite article for both types of names. In either case, no marking difference and thus no markedness asymmetry can be determined.

In sum, the different proper name types often do confirm the proposed markedness hierarchy at least with respect to definiteness marking, but there are also many deviations. It is obvious that more data are necessary to approach a solution to these questions.

The question remains: are there other functional or pragmatic explanations for the use of the definite articles with proper names beyond the one Van Langendonck proposed for the markedness asymmetries in proper names. I think there are at least four functional motivations that should be considered in the future research on this question.

First, the definite article may be used with personal names, because they express some kind of emotive function. In the spoken varieties of Northern German, the use of definite articles with personal names has a negative connotation, and this is reported from other languages too (in case that there is a choice, of course).

Second, the definite article may acquire a kind of classifying function with regard to the different name types. This kind of exaptation of the definite article has been dealt with extensively in Nübling (2020). Compare the two examples from English and German below.

(51) English
 Queen Elizabeth (person) vs. ***The** Queen Elisabeth* (ship)

(52) German
 Ø Mercedes (female person)
 der (DEF.M) *Mercedes* (car)
 das (DEF.N) *Mercedes* (hotel or restaurant)
 die (DEF.F) *Mercedes* (ship or motorcycle)

In English, the usage of the definite article with a female personal name converts this name into a name of a ship. In German, the definite article indicating different genders converts a female personal name into a name for a car, for a hotel or restaurant, or for a ship.

Third, the definite article marks indeed definiteness with common nouns (appellative expressions) that are meant and used as personal names by the speaker. The functional motivation for this could be the need to clearly distinguish common nouns or more generally referential phrases with common nouns from proper names. This can be illustrated with examples from German. Among German toponyms, there are many that are etymologically more or less transparent. Compare the following examples (see Nübling, this volume).

(53) *der   Schwarzwald*
     der   schwarz-wald
     DEF.M   black-forest
     'Black Forest'

(54) *der   Bodensee*
    der   boden-see
    DEF.M ground-lake
    'Lake Constance'

(55) *die   Zugspitze*
    die   zug-spitze
    DEF.F avalanche.path-summit
    'Zugspitze (no Engl. equivalent)'

They all take definite articles. But personal names, in particular first names, and settlement names and other place names are etymologically opaque. That is, these opaque forms are lexicalized as proper names. They can be recognized instantly as proper names by the speakers, because they are part of the Onomasticon of the language. Compare the personal and place names below.

(56) First names: *Thomas, Paul, Petra, Simone*, etc.
    Settlement names: *München, Köln, Passau*, etc.

One could propose the following hypotheses:
a) The more proper names are formally (phonologically) dissociated from the corresponding appellative forms, the weaker the need to mark its proper name status grammatically.
b) The higher proper names are on the hierarchies postulated above, the more likely it is that they are formally/phonologically dissociated from the corresponding appellative expressions.
c) The higher the proper names are on the hierarchies postulated above, the higher their frequency of use in the speech community, and the less the need to mark them specifically.

Fourth, the definite article may be grammaticalized as a general proper name marker. The need to distinguish referential phrases with common nouns and with proper names becomes a general grammatical rule. This seems likely for the languages that take obligatorily the definite article with proper names such as Greek, Hidatsa, and Seri. Note that in Hidatsa, only native personal names take the definite article. Names borrowed from English don't, perhaps because they are lexicalized proper names and as such easily identifiable as names.

In conclusion, it seems that the question what factors the co-occurrence of proper names and definite articles motivate is still not solved. There are probably different factors or motivations that trigger or control the definite article to occur with personal names, with place names, and with the other name types that were not treated in this survey.

Abbreviations

1, 2, 3	first, second, third person
3R	third person realis
ABL	ablative
ACC	accusative
ALL	allative
AOR	aorist
ART(.ID)	article
ASSOC	associative plural
COM	comitative
CONJ	conjunction
CORE	article for core arguments
CRD	coordinate conjunction
DAT	dative
DEB	debitive
DECL	declarative
DEF	definiteness marker/definite article
DES	desiderative
DET	determiner
DIR	directional
EMPH	emphatic
ERG	ergative
EXS	existential
F	feminine
FIRSTN	first name
FUT	future tense
GEN	genitive
INF	infinitive
INST	instrumental
LASTN	last name
LOC	locative

M	masculine
N	neuter
NC	noun class
NICKN	nickname
NOM	nominative
NP	noun phrase
OBJ	object
PERSN	personal name
PF	perfect
PL	plural
PN	proper name
POSS	possession
PRES	present tense
PST	past tense
PTCP	participle
RECIP	reciprocal
RLZ	realization
RP	referential phrase
SBJ	subject
SG	singular
SUB	subordinator

Acknowledgements

I am grateful to Katharina Eberwein, Sarah Thanner, Corinna Handschuh, and Sandra Johne for helping me to collect the relevant data for this survey. Many thanks go also to the two reviewers of this paper for the numerous suggestions for improvement and corrections. Last but not least, I would like to thank Maximilian Weiss for helping with the layout and last corrections.

References

Ahland, Colleen Anne. 2012. *A grammar of Northern and Southern Gumuz*. Eugene, OR: University of Oregon PhD thesis.

Arkoh, Ruby & Lisa Matthewson. 2013. A familiar definite article in Akan. *Lingua* 123. 1–30.

Bellmann, Günther. 1990. *Pronomen und Korrektur. Zur Pragmalinguistik der persönlichen Referenzformen*. Berlin & New York: Walter de Gruyter.
Berchtold, Simone & Antje Dammel. 2014. Kombinatorik von Artikel, Ruf- und Familiennamen in Varietäten des Deutschen. In Friedhelm Debus, Rita Heuser & Damaris Nübling (eds.), *Linguistik der Familiennamen* (Germanistische Linguistik 225–227), 249–280.
Besnier, Niko. 2000. *Tuvaluan. A Polynesian language of the Central Pacific*. London: Routledge.
Carlson, Robert. 1994. *A grammar of Supyire*. Berlin & New York: Mouton de Gruyter.
Caro Reina, Javier. 2020. The definite article with place names in Romance languages. In Nataliya Levkovych & Julia Nintemann (eds.) *Aspects of the grammar of names: Empirical case studies and theoretical topics* (LINCOM Studies in Language Typology 33), 25–53. München: LINCOM.
Cunha, Celso Ferreira da & Luís Filipe Cintra. 1984. *Nova gramática do português contemporâneo*. Lisboa: Sá da Costa.
Croft, William. 2001. *Radical construction grammar. Syntactic theory in typological perspective*. Oxford: Oxford University Press.
Croft, William. 2016. Comparative concepts and language-specific categories: Theory and practice. *Linguistic Typology* 20(2). 377–393.
Cumberland, Linda A. 2005. *A Grammar of Assiniboine: A Siouan language of the Northern Plains*. Bloomington, IN: Indiana University PhD thesis.
Donohue, Mark. 1999. *A grammar of Tukang Besi*. Berlin & New York: Mouton de Gruyter.
Dryer, Matthew S. 2013. Definite articles. In Matthew S. Dryer & Martin Haspelmath (eds.), *The world atlas of language structures online*. Leipzig: Max Planck Institute for Evolutionary Anthropology. http://wals.info/chapter/37 (checked 09.11.2017).
Dum-Tragut, Jasime. 2009. *Armenian. Modern Eastern Armenian*. Amsterdam & Philadelphia: John Benjamins.
Edgar, John. 1989. *A Masalit grammar*. Berlin: Dietrich Reimer.
Fedden, Sebastian. 2011. *A Grammar of Mian*. Berlin: Mouton de Gruyter.
Haspelmath, Martin. 2010. Comparative concepts and descriptive categories in crosslinguistic studies. *Language* 86(3). 663–687.
Hawkins, John A. 1978. *Definiteness and indefiniteness. A study in reference and grammaticality prediction*. London: Croom Helm.
Heath, Jeffrey. 1999. *A Grammar of Koyra Chiini. The Songhay of Timbuktu*. Berlin & New York: Mouton de Gruyter.
Heath, Jeffrey. 2008. *A Grammar of Jamsay*. Berlin: Mouton de Gruyter.
Heidenkummer, Alexandra & Johannes Helmbrecht. 2017. Form, Funktion und Grammatikalisierung des Eigennamenmarkers =ga im Hoocąk (Sioux). In Johannes Helmbrecht, Damaris Nübling & Barbara Schlücker (eds.). *Namengrammatik* (Linguistische Berichte – Sonderheft 23), 11–33. Hamburg: Buske.
Hellwig, Birgit. 2011. *A Grammar of Goemai*. Berlin & Boston: Mouton de Gruyter.
Hellwig, Birgit. 2019. *A Grammar of Qaqet*. Berlin & New York: Mouton de Grutyer.
Helmbrecht, Johannes. 2020. On the morphosyntax of personal names in Hoocąk (Sioux). In Nataliya Levkovych & Julia Nintemann (eds.), *Aspects of the grammar of names: Empirical case studies and theoretical topics* (LINCOM Studies in Language Typology 33), 147–167. München: LINCOM.
Himmelmann, Nikolaus P. 1997. *Deiktikon, Artikel, Nominalphrase. Zur Emergenz syntaktischer Struktur*. Tübingen: Niemeyer.

Jany, Carmen. 2009. *Chimariko grammar: Areal and typological perspective.* Los Angeles: University of California Press.
Kenesei, István, Robert M. Vago & Anna Fenyvesi. 1998. *Hungarian.* London & New York. Routledge.
König, Ekkehard. 2018. Definite articles and their uses. In Daniël Olmen, Tanja Mortelmans & Frank Brisard (eds.), *Aspects of linguistic variation*, 165–184. Berlin & Boston: De Gruyter.
Kripke, Saul. 1980. *Naming and necessity.* 2nd edn. Oxford: Blackwell.
Le Bot, Marie-Claude & Martine Schuwer. 2015. Remarques sur la morphosyntaxe des toponymes complexe en français. In Jonas Löfström & Betina Schnabel-Le Corre (eds.), *Challenges in synchronic toponymy*, 203–218. Tübingen: Narr.
Lepschy, Anna Laura & Giulio Lepschy. 1986. *Die italienische Sprache. Mit einem Vorwort von Jörn Albrecht.* Tübingen: Francke.
Löbner, Sebastian. 1985. Definites. *Journal of Semantics* 4. 279–326.
Lyons, Christopher. 1999. *Definiteness.* Cambridge: Cambridge University Press.
Mahootian, Shahrzad. 1997. *Persian.* London: Routledge.
Maiden, Martin & Cecilia Robustelli. 2000. *A reference grammar of Modern Italian.* London: Arnold.
Marfo, Charles. 2005. *Aspects of Akan grammar and the phonology-syntax interface.* Hong Kong: Lambert Academic Publishing.
Marlett, Stephen A. 2008. The form and use of names in Seri. *International Journal of American Linguistics* 74(1). 47–82.
Matasović, Ranko. 2010. *A short grammar of East Circassian* (Kabardian). University of Zagreb. http://mudrac.ffzg.unizg.hr/~rmatasov/KabardianGrammar.pdf (checked 11/11/17).
McGregor, William. 1996. *Nyulnyul.* München: LINCOM Europa.
Nübling, Damaris, Fabian Fahlbusch & Rita Heuser. 2015. *Namen. Eine Einführung in die Onomastik.* 2nd edn. Tübingen: Narr.
Nübling, Damaris. 2020. *Die Capital – der Astra – das Adler*: The emergence of a classifier system for proper names in German. In Renata Szczepaniak & Johanna Flick (eds.), *Walking on the grammaticalization path of the definite article* (Studies in Language Variation 23), 228–249. Amsterdam & Philadelphia: John Benjamins.
Obata, Kazuko. 2003. *A grammar of Bilua. A Papuan language of the Solomon Islands.* Canberra: Pacific Linguistics.
Park, Indrek. 2012. *A grammar of Hidatsa.* Indiana University PhD thesis.
Press, Ian. 1986. *A grammar of Modern Breton.* Berlin: De Gruyter.
Schuh, Russell G. 1998. *A grammar of Miya.* Berkeley: University of California Press.
Seyoum, Mulugeta. 2008. *A grammar of Dime.* Utrecht: LOT.
Soukka, Maria. 2000. *A descriptive grammar of Noon. A Cangin language of Senegal* (LINCOM Studies in African Linguistics 40). München: LINCOM Europa.
Stirtz, Timothy M. 2011. *A grammar of Gaahmg. A Nilo-Saharan language of Sudan.* Utrecht: LOT.
Stolz, Thomas, Nataliya Levkovych & Aina Urdze. 2017. Die Grammatik der Toponyme als typologisches Forschungsfeld: eine Pilotstudie. In: Johannes Helmbrecht, Damaris Nübling & Barbara Schlücker (eds.), *Namengrammatik* (Linguistische Berichte – Sonderheft 23), 121–146. Hamburg: Buske,.
Stolz, Thomas, Nataliya Levkovych & Aina Urdze. 2018. La morfosintassi dei toponimi in prospettiva tipologica. In Giuseppe Brincat & Sandro Caruana (eds.), *Tipologia e 'dintorni': il metodo tipologico alla intersezione di piani d'analisi*, 307–324. Roma: Bulzoni.

Stolz, Thomas & Nataliya Levkovych. 2020. Grammatical versus onymic classifiers: First thoughts about a potentially interesting topic. In Nataliya Levkovych & Julia Nintemann (eds.), *Aspects of the grammar of names: Empirical case studies and theoretical topics* (LINCOM Studies in Language Typology 33), 167–180. München: LINCOM.

Stolz, Thomas & Nataliya Levkovych. this volume. On Special Onymic Grammar (SOG) in Fijian and related languages.

Terrill, Angela. 2003. *A grammar of Lavukaleve*. Berlin: Mouton de Gruyter.

Van Langendonck, Willy. 2007. *Theory and typology of proper names*. Berlin & New York: Mouton de Gruyter.

Wegener, Claudia. 2012. *A grammar of Savosavo*. Berlin & Boston: Walter de Gruyter.

Werth, Alexander. 2014. Die Funktionen des Artikels bei Personennamen im norddeutschen Sprachraum. In Friedhelm Debus, Rita Heuser & Damaris Nübling (eds.), *Linguistik der Familiennamen* (Germanistische Linguistik 225–227), 139–174. Hildesheim: Olms.

Werth, Alexander. 2020. *Morphosyntax und Pragmatik in Konkurrenz. Der Definitartikel bei Personennamen in den regionalen und historischen Varietäten des Deutschen* (Studia Linguistica Germanica 136). Berlin & Boston: Walter de Gruyter.

Willett, Thomas L. 1991. *A reference grammar of Southeastern Tepehuan*. Dallas: Summer Institute of Linguistics & UT-Arlington.

Wotjak, Gerd. 1985. Zur Semantik der Eigennamen. *Namenkundliche Informationen* 48. 1–17.

Appendix

#	Languages	Geographical area	Genetic affiliation	References
1.	Akan	Ghana	Niger-Congo, Kwa	Marfo (2005), Arkoh and Matthewson (2013)
2.	Arabic, Standard	various countries	Afro-Asiatic	Author (p.c.)
3.	Armenian, Modern	Armenia	Indo-European, Armenian	Dum-Tragut (2009)
4.	Assiniboine	Canada, United States	Siouan	Cumberland (2005)
5.	Bilua	Solomon Islands	Solomons East Papuan, Bilua	Obata (2003)
6.	Breton, Modern	France	Indo-European, Celtic	Press (1986)
7.	Bulgarian	Bulgaria	Indo-European, Slavic	Author (p.c.)
8.	Chimariko	United States	Isolate of California	Jany (2009)
9.	Dime	Ethiopia	Afro-Asiatic, South Omotic	Seyoum (2008)
10.	English	United King-	Indo-European, Ger-	Author (p.c.)

#	Languages	Geographical area	Genetic affiliation	References
		dom, Ireland	manic	
11.	French	Switzerland, France	Indo-European, Romance	Author (p.c.), and Le Bot and Schuwer (2015)
12.	German	Germany, Austria, Switzerland	Indo-European, Germanic	Nübling et al. (2015)
13.	German, Coll.	Germany	Indo-European, Germanic	Author (p.c.)
14.	Goemai	Nigeria	Afro-Asiatic, West Chadic	Hellwig (2011)
15.	Greek, Modern	Greece	Indo-European, Greek	Author (p.c.)
16.	Gumuz	Ethiopia, Sudan	Gumuz	Ahland (2012)
17.	Hidatsa	United States	Siouan	Park (2012)
18.	Hoocąk	United States	Siouan	Helmbrecht (2020), Heidenkummer and Helmbrecht (2017)
19.	Hungarian	Hungary	Uralic	Kenesei, Vago and Fenyvesi (1998)
20.	Ingessana	Sudan	Eastern Sudanic, Eastern Jebel	Stirtz (2011)
21.	Italian	Italy, Switzerland	Indo-European, Romance	Maiden and Robustelli (2000), Lepschy and Lepschy (1986)
22.	Italian, Coll.			Author (p.c.)
23.	Jamsay	Mali	Dogon	Heath (2008)
24.	Kabardian	Russia	Abkhaz-Adyge	Matasović (2011)
25.	Koyra Chiini	Mali	Songhay	Heath (1999)
26.	Lavukaleve	Solomon Islands	Solomons East Papuan, Lavukaleve	Terrill (2003)
27.	Lepcha	Bhutan, India, Nepal	Sino-Tibetan, Lepcha	Plaisir (2006)
28.	Masalit	Chad, Sudan	Maban	Edgar (1989)
29.	Mian	Papua New Guinea	Trans-New Guinea, Ok	Fedden (2011)
30.	Miya	Nigeria	Afro-Asiatic, West Chadic	Schuh (1998)

#	Languages	Geographical area	Genetic affiliation	References
31.	Noon	Senegal	Niger-Congo, Northern Atlantic	Soukka (2000)
32.	Nuuchahnulth	Canada	Wakash	Inman (p.c.)
33.	Nyulnyul	Australia	Nyulnyulan	McGregor (1996)
34.	Persian	Iran	Indo-European, Indo-Iranian	Mahootian (1997)
35.	Qaqet	Papua New Guinea	Baining	Hellwig (2019)
36.	Savosavo	Solomon Islands	Solomons East Papuan, Savosavo	Wegener (2012)
37.	Seri	Mexico	Isolate of Mexico	Marlett (2008)
38.	Slovene	Slovenia	Indo-European, Slavic	Author (p.c.)
39.	Spanish	Spain	Indo-European, Romance	Author (p.c.)
40.	Supyire	Mali	Niger-Congo, Gur	Carlson (1994)
41.	Tepehuan	Mexico	Uto-Aztecan	Willett (1991)
42.	Tuavaluan	Tuvalu	Austronesian, Oceanic	Besnier (2000)
43.	Tukang Besi	Indonesia	Austronesian, Celebic	Donohue (1999)

Elisheva Jeffay and Susan Rothstein
On personal names in construct states in Modern and Biblical Hebrew

Abstract: In this paper, we explore the contrasts between the distribution of personal names in Modern and Biblical Hebrew, including (i) the lexical semantics of proper names in Biblical Hebrew, (ii) rare examples of proper names with definite articles, (iii) place names, which may have the syntax and morphology of common nouns, (iv) uses of proper names as sentential predicates. We suggest, following Longobardi (1994), that there is a change in the predicate vs. argument status of proper names, and note the absence of explicit morphosyntactic evidence for determiners in Biblical Hebrew, suggesting that proper names could only interpreted as being of the NP type.

Keywords: construct state, DPs, Hebrew, NP predicates, personal names.

1 Introduction

The construct state is a grammatical construction that exists in a number of languages, most significantly Semitic languages, and is often used to express possessive relations, with comparisons drawn between its semantic use and that of the genitive in other languages. Much research has been done on the construct phrase in Hebrew, including Borer (2009), who draws a clear line between compounds and construct phrases, defining the construct phrase as that which is semantically built up from its individual syntactic parts. The construct phrase consists of a head, generally nominal, and an annex, often referred to as a dependent by typologists, which is not necessarily so, and may denote a property or feature of the head. In Modern Hebrew, Rothstein (2012, 2017) notes that the distribution of personal names is severely restricted in the annex position of construct phrases, to the extent that it could be reasonably claimed that proper names cannot appear in this position. In contrast, proper names are able to freely appear in the annex position in Biblical Hebrew. Therefore, the construct in (1), with a definite NP (Noun Phrase) in the annex posi-

---

Elisheva Jeffay: Bar Ilan University, Ramat Gan, 5290002, emjeffay@gmail.com
Susan Rothstein: Bar Ilan University, Ramat Gan, 5290002

https://doi.org/10.1515/9783110672626-006

tion, is felicitous both in Modern and Biblical Hebrew (with a few minor phonological changes), while (2), with a proper name in annex position, is grammatical only in Biblical Hebrew.[1]

(1) *bet   ha-sar*
    house.CS   DEF-official
    'the official's house'

(2) *bəney   ya'aqoḇ*   (Gen 34:13)
    son.M.PL.CS Jacob
    'the sons of Jacob'

The construct phrase is not the only possessive construction used in Modern Hebrew, where the šel-genitive is common, nor in Biblical Hebrew, where other expressions of possession such as the pronominal suffix are used. In this paper, we will focus on the construct phrase, which we will use in order to demonstrate the contrast in distribution of proper names between Modern and Biblical Hebrew. We will argue in this paper that the distribution of proper names in Biblical Hebrew, their use as appellatives and their appearance in conjunction with the definite article strongly suggest that personal names are NP predicates, i.e. noun phrases that denote properties and can be used in certain situations as modifiers, in Biblical Hebrew. This is where Modern Hebrew differs: We suggest that personal names are referential in Modern Hebrew, and therefore form DPs (Determiner Phrases) rather than NPs (there is reason to think that Biblical Hebrew does not have a DP position). DPs are determined, i.e. closed to further modification or definiteness, and are generally referential in nature. We argue that the structure of the construct state is the same in Biblical and Modern Hebrew, and this means that in both, the annex position is an NP position and not a DP position. In Modern Hebrew, therefore, the inability of proper names to occur in the annex position follows from this.

Section 2 of this paper will outline the Modern Hebrew data, which will act as a starting point for the analysis, as the closest language to Biblical Hebrew for which a theory has been developed about the construct state. In Section 3, we describe the Biblical Hebrew data and show how it contrasts with the Mod-

---

[1] The quotations from the Biblical text will be taken from the standard Hebrew Bible, as found on the Accordance Software which has been used for all quotations in this paper, with transliterations according to the standard protocol among Biblical Hebrew scholars. This will therefore appear to differ from the Modern Hebrew even in situations in which the Hebrew would be written the same.

ern Hebrew data; we argue that there is additional evidence in support of an analysis of Biblical Hebrew names as NPs; and we propose an analysis of the data in line with Longobardi (1994). In Section 4, we discuss the potential theoretical implications.

2 The Modern Hebrew data

In this section, we will explore the Modern Hebrew data, explaining how the construct phrase works, and the constraints on it that prevent the use of personal names as annexes of the phrase, as well as the few exceptions to these rules.

We point out that we work with the assumption that, unlike referential proper names, referential definite noun phrases in Modern Hebrew *can* be predicative NPs, rather than DPs. In English, the DP *the teacher* would be interpreted as σ(TEACHER), which denotes the unique teacher if there is a unique teacher in the given context and is otherwise undefined. In Modern Hebrew, we can assume that the predicative NP *ha-mora* denotes {σ(TEACHER)}, the singleton set that contains the unique teacher in the context in which there exists a unique teacher. The referential effect found in R-constructs is derived by letting the modification relation be "being possessed by the unique individual in the interpretation of the NP annex". Our assumption is that this analysis is not possible for referential proper names in Modern Hebrew because they are DPs and the annex cannot be a DP.

2.1 The construct

In this section, we outline the properties of the construct phrase in Modern Hebrew, and in the following subsections, we will expand our study to include the distribution of personal names in relation to the construct, and then outline some additional evidence for the referential nature of personal names in Modern Hebrew.

There are many possessive constructions cross-linguistically, but we bring here an example from English, in which the most common possessive constructions are the free genitive, where the possessive relation is expressed via the preposition *of* (as in (3a)), and the Saxon genitive, where the possessor is marked morphologically by genitive case (as in (3b)).

(3) a. *The house of John*
     b. *John's House*

Having demonstrated what we are referring to, we move to the Semitic languages under discussion. In Semitic languages, there is a possessive construction in which the head, denoting the possessum, is marked morphologically by *construct* morphology. This may include a shortening of the final vowel, insertion of a consonant when the head ends with a vowel, or in some circumstances, no morphological markings at all. Explicit construct morphology is demonstrated in the following example from Modern Standard Arabic:

(4) *sayyāra-t-u   l-rağul-i* (Kremers 2003: 48)
     car.F-CS-NOM   DEF-man-GEN
     'the man's car'

In (4), the head is marked as being in the construct state, while the possessor is marked by the genitive, as is possible in Arabic. This demonstrates that construct morphology does not necessarily replace the genitive, as well as that the construct is not interchangeable with the genitive as, in this phrase, there is double marking, in which the possessor and possessum both host different markers with the same function, i.e. denoting the possession relationship. In the case of the possessor this is the genitive, while in the case of the possessum it is the construct state.

Modern Hebrew has (at least) two genitive constructions in which possession can be expressed. As is the case in English, Modern Hebrew has a free genitive, in which the possessum is the head of the phrase, in the unmarked absolute state, and the possessor is the object of the preposition *šel* 'of' (as in (5)).

(5) *ha-bayit   šel   ha-mora*
     DEF-house.M.SG.ABS   of   DEF-teacher.F.SG.ABS
     'the teacher's house'

Additionally, like in Arabic, there is the *construct*, in which the possessum, the head of the phrase, is in the construct state, and the possessor is in the unmarked absolute state, acting as the direct complement of the head (as in (6)). This position is referred to as the *annex* of the construct phrase. There is no explicit genitive case.

(6) bet ha-mora
 house.M.SG.CS DEF-teacher.F.SG.ABS
 'the teacher's house'

A number of Biblical Hebrew scholars, such as Gesenius (GKC 1910), argue that pre-Biblical Hebrew may well have had an explicit genitive case, as is found in other Semitic languages, but that it has not survived into Biblical Hebrew. There is no genitive case in either Biblical or Modern Hebrew, and Modern Hebrew primarily uses either the free genitive or the construct, as outlined in the next subsection.

2.2 Properties of constructs

There are a number of important properties which the construct phrase has, both morphologically and lexically. Most clearly, heads in the construct state often differ morphologically from non-construct state heads of phrases, in part due to stress reduction as in (7a). This is sometimes a useful diagnostic for recognising construct phrases.

(7) a. *bayit* 'house' → *bet*
 b. *mənora* 'candelabra' → *mənorat*

There are a number of important syntactic properties. No lexical item is permitted to intervene between the head and its genitive (i.e. the annex of the phrase), and any modifiers of the head must follow the annex. This is shown in the contrast between (8a) and (8b), where *ha-mora* is the annex:

(8) a. bet ha-mora he-xadaš
 house.M.SG.CS DEF-teacher.F.SG.ABS DEF-new.M.SG
 'the teacher's new house'
 b. #bet he-xadaš ha-mora
 house.M.SG.CS DEF-new.M.SG DEF-teacher.F.SG.ABS
 'the teacher's new house'

There is no indefinite marker in Hebrew: Indefiniteness is expressed simply by the lack of a definite marker. Definiteness is not marked on the head, but on the annex, and percolates down to the whole phrase. Therefore, the lack of the definiteness marker on the head *bet* in (8a) does not indicate that the phrase is indefinite, as the marker is present on the annex, and the phrase denotes a specific new house of the teacher; it does not mean "a house of the new teacher".

Borer (2009) outlines two different types of construct phrases, *modificational* and *referential*, which have different roles and vary slightly in terms of what they can take as an annex.

2.2.1 Referential constructs (R-constructs)

Referential constructs express a relation between the head and an apparently referential annex. In these construct phrases, the annex is usually definite. These constructs can denote a range of relationships, including possession (as in (9a)), part-whole relations (as in (9b)), and group membership (as in (9c)).

(9)   a.   *bet*   *ha-mora*
               house.M.SG.CS   DEF-teacher.F.SG.ABS
               'the teacher's house'
        b.   *regel*   *ha-šulxan*
               foot.F.SG.CS   DEF-table.M.SG.ABS
               'the foot of the table'
        c.   *xaver*   *ha-va'ada*
               member.M.SG.CS DEF-committee.F.SG.ABS
               '(the) member of the committee'

2.2.2 Modificational constructs (M-constructs)

In modificational constructs, the annex expresses a property of the head. Although the annex is a noun, it has a "quasi-adjectival" function. In such constructs, a range of modificational relations are available, including matter (as in (10a)), purpose (as in (10b)), and kind (as in (10c)).

(10)   a.   *bet*   *(ha-)ec*
               house.M.SG.CS   DEF-wood.M.SG.ABS
               'the/a wooden house'
        b.   *magev-et*   *(ha-)mitbax*
               towel-F.SG.CS   DEF-kitchen.M.SG.ABS
               'the/(a) kitchen towel'
        c.   *xatul-at*   *(ha-)bayit*
               cat   -F.SG.CS   DEF-house.M.SG.ABS
               'the/(a) domestic cat'

Modificational constructs have the same signature properties as referential constructs, particularly with definiteness expressed on the annex but interpreted on the whole phrase. These examples of the modificational use of the construct demonstrate that the construct is not merely a vehicle for the expression of possession. Modificational constructs are becoming the dominant kind of construct used in contemporary Modern Hebrew.

In the next subsection, we will discuss distribution of personal names within it and draw some conclusions as to why the distribution is this way.

2.3 Personal names in construct phrases

Personal names do not occur with a referential interpretation in the annex of R-constructs in Modern Hebrew. This is surprising, since personal names denote individuals, which are the quintessential possessors.

(11)   a. *halax-ti*   *le-bet*   *ha-mora*
           went-I   to-house.M.SG.CS   DEF-teacher.F.SG.ABS
           'I went to the teacher's house.'
   b. #*halax-ti* *le-bet*   *ariela*
           went-I   to-house.M.SG.CS   Ariella
           Intended: 'I went to Ariella's house.'

(11b) is infelicitous if *ariela* is interpreted as a referential proper name. There is, however, an alternative use of the expression which is felicitous, namely where *ariela* is not referential, and *bet ariela* is used as the name of the public library in Tel Aviv. Crucially, this would not be translated into English as *Ariella's library*, but rather as *The Ariella Library*, showing that indeed, *ariela* is not referential in the felicitous interpretation.

Personal names can also form the annex of M-constructs, in which they, like common nouns, act quasi-adjectivally, denoting a type that is readily identifiable and in no way context dependent (Danon 2017).

(12)   *miškef-ey*   *jon lenon*
        glass(es)-M.PL.CS   John Lennon
        'John Lennon-style glasses' (not: 'glasses which were owned by John Lennon')

Rothstein (2012, 2017) discusses this data and argues that the annex of a construct in Hebrew is both in R-constructs and M-constructs a *predicate* NP, that is, an NP with a predicative interpretation. If we can argue successfully that referential personal names in Hebrew must be DPs and cannot be predicative NPs, we can explain why they cannot occur in annex position.

In English, personal names such as *John* allow for uses where they are clearly interpreted as NPs, although differently to the kind interpretation that is found in (12). (13a) shows an example of this, where the proper names occur as the complement of a numerical determiner and allows number morphology as occurs with NPs, while 13b shows an example where the proper name occurs as the complement of the definite article and allows for modification via a restrictive relative clause, like NPs do.

(13)   a. There are **three** *Johns* in the class
        b. **The** *John* **I know** would never have spoken like that!

However, the examples here in (13) are closer to the referential uses of proper names than the example in (12), since the three *Johns* in (13a) are three specific individuals with the name *John*, and the *John* I know in (13b) is indeed called John.

We find neither of these uses in Modern Hebrew, as shown in (14):

(14)   a. #yeš   šloša   david-im   b-a-kita
           there are three   David-M.PL in-DEF-class
           Intended: 'There are three Davids in the class.'
    b. #ha-david še-ani makira lo   haya medaber kaxa
          DEF-David that-I know   not AUX speak   so
          Intended: 'The David that I know wouldn't speak like that.'

As can be seen, the personal name *david* does not allow the NP referential use that is possible in the English examples found in (13). A reasonable explanation for this is that referential names in Modern Hebrew must therefore be DPs. DPs, while also referential, are closed to further modification such as number or definiteness, while NPs would be open to such modification.

Primarily, in Modern Hebrew, place names form an exception to the above observations, in that they seem to be a type of referential proper names that is allowed in the annex of constructs, which is not the case with proper names, considered to be the typical example of referential names, as shown in (15a) and (15b):

(15) a. *iriyat*   *nahariya*
         town council.F.SG.CS   Nahariya
         'the Nahariya town council'
      b. *tošvey*   *tel aviv*
         resident.M.PL.CS   Tel Aviv
         '(the) residents of Tel Aviv'

There is reason to think that, unlike personal names, place names in Modern Hebrew are indeed NPs, although further research will be required to understand why they did not make the shift that personal names did. Unlike personal names, place names can occur as the complement of the definite article and allow for modification with restrictive relative clauses, as in (16):

(16) *ha-tel aviv   še-ani   ohevet*
      DEF-Tel Aviv   that-I   love
      'The Tel Aviv that I love.'

Place names can also be the complement of quantifiers like *kol-* 'every':

(17) *b-a-mirpeset   ha-madhima*
      in-DEF-balcony.F.SG.ABS   DEF-wonderful.F.SG
      še-maškifa   al kol tel aviv
      that-looks out   on all Tel Aviv
      'In the wonderful balcony that looks out on all Tel Aviv.'

In Biblical Hebrew, it is to be expected that place names would also have the properties of NP predicates, given that proper names in general have these properties. This is the case, as demonstrated in (18), in which the Jordan river is the annex of the construct phrase which is then referred to by a quantifier:

(18) *'eṭ-kol-kikkar   ha-yyarden*
      DOM-all-plain.CS   DEF-Jordan
      'All of the Jordan plain'

As can be seen, place names differ from proper names as used to denote people. They are able to act as the annex of the construct phrase while maintaining their reference to the place they denote and can be modified in ways that are disallowed for Modern Hebrew *personal* proper names, indicating strongly that there is a significant difference between the two groups of names. Thus, there is good reason to assume that place names did not shift to DPs in Modern Hebrew but remained as predicative NPs.

We have explored the Modern Hebrew data, and in the next section, we will outline the Biblical Hebrew data, showing the contrasts between it and the data taken from Modern Hebrew.

3 The construct phrase in Biblical Hebrew

In this section, we will examine the data that refers to the construct phrase and the personal proper name. We will explore the contrasts between Modern and Biblical Hebrew in terms of the personal proper name in the annex of the construct phrase, as well as the evidence for the personal proper name having a different status in Biblical and Modern Hebrew.

3.1 The Biblical Hebrew data

Biblical Hebrew differs from Modern Hebrew in that there is no free genitive construction in Biblical Hebrew, while there is in Modern Hebrew, as we saw in Section 1. There is another structure in Biblical Hebrew which can express possession, namely, structures with a relative marker:

(19) ha-c'on ašer le-'avi-ha (Gen 29:9)
 DEF-sheep REL to-father-POSS.3.F.SG
 'her father's sheep' (lit. 'the sheep which were to her father')

In Biblical Hebrew, the construct state looks similar in terms of its morphology to the construct state in Modern Hebrew. (19) is grammatical as a construct phrase in both Modern and Biblical Hebrew, and referential constructs such as (19) and modificational constructs such as (20) are common in Biblical Hebrew:

(20) beyt ha-melek (2Sam. 11:2)
 house.M.SG.CS DEF-king.M.SG.ABS
 'the king's house'
(21) mənorat ha-zahav (2Chr. 13:11)
 candelabra.F.SG.CS DEF-gold.M.SG.ABS
 'the candelabra of gold'

The crucial difference, however, is that in contrast to Modern Hebrew, personal names appear freely in the annex of R-constructs in Biblical Hebrew. The proper name R-construct can denote a range of relationships, including kinship relations (as in (22a)), possession or control (as in (22b) and (22c)), and inalienable possession or part-of relations (as in (22d)).

(22) a. *bəney   ya'aqob*   (Gen 34:13)
        son.M.PL.CS   Jacob
        'the sons of Jacob'
     b. *beyt   mika̠*   (Judg. 17:8)
        house.M.SG.CS   Micha
        'the house of Micha'
     c. *bə-'aholey   šem*   (Gen. 9:27)
        in-tent.M.PL.CS Shem
        'in the tents of Shem'
     d. *bə`-'eyney   avraham*   (Gen. 21:11)
        in-eye.M.PL.CS   Abraham
        'in the eyes of Abraham'

Taking all of this into account, a clear contrast can be seen between the Biblical and Modern Hebrew data. Why is it that personal names can be found in the annex of construct phrases in Biblical Hebrew when they cannot be found in this position in Modern Hebrew?

There are two strategies for explaining the difference. The first is to assume that the construct phrase in Biblical Hebrew is different from that in Modern Hebrew, for instance, in that the annex position can take referential DPs. The second strategy is to assume that personal names in Biblical Hebrew are predicates, and are therefore able to take the annex position, being incorporated into the construct phrase via NP-incorporation. In this paper, we will defend the second of these strategies, and point out a number of pieces of evidence supporting this option over the first.

Assuming that Modern Hebrew put a constraint on the construct disallowing DP annexes that was not present in Biblical Hebrew would be rather surprising from a diachronic point of view. Generally speaking, constraints on the construct have stayed relatively stable from Biblical Hebrew through to Modern Hebrew except for in regard to proper names. If anything, we find that constraints have relaxed, for example, with constructs becoming input to lexicalisation processes. This is something that is extremely rare, if not impossible, in Biblical Hebrew, but becomes increasingly common from Rabbinic Hebrew (ca. 1st-5th century CE onwards) (Schniedewind 2013: 194). Additionally, constructs headed by a range of numerical classifiers and quantity expressions become more frequent moving into Modern Hebrew. The restriction of NPs only as annexes of the construct would go in the opposite direction of this general tendency.

We will not take this view, as we think that there is additional evidence that proper names are predicates in Biblical Hebrew, as is shown in Section 3.2 below.

3.2 Additional evidence that proper names are predicates in Biblical Hebrew

3.2.1 Lexical argument: Proper names as appellatives

One piece of evidence in support of the predicative nature of personal names in Biblical Hebrew is the following quotation from Hess (2013): "[personal names] are most often composed of one or more lexical items that can be readily identified as part of the inventory of common nouns, verbs, and other lexical items in their originating language". Examples of this include *qorax* 'bald one' and *adoniy-a* 'my Lord is Y'ah'.

In contrast to other Hebrew nouns, proper names are frequently non-templatic and may involve compounds expressing relevant properties of the person named (GKC 1910: §81d)

(23)   a. *yeho-natan (= Jonathan)*   (1Sam 14:6)
         Ya-gave
         i.e. 'given by God'
    b. *obad-ya (= Obadiah)*   (Obad 1:1)
         servant-God
         i.e. 'servant of God'
    c. *mo'ab (= Moab, a child of incest)*   (Gen 19:37)
         from-father
         i.e. 'from father'

This suggests that many personal names, and indeed names in general, began as *appellatives* (GKC 1910: §125d–f), which are NPs used to characterise a given individual via a highly salient characteristic. It is reasonable to assume that an appellative predicate, which characterises a *given* individual, denotes a singleton set. We would therefore expect appellative predicates to pattern with definite predicative NPs.

We observe that many proper names in Biblical Hebrew have origins in appellative predicates, and it is reasonable to assume that appellative predicates pattern with definite predicative NPs. We therefore assume that the proper names in question preserve that semantic interpretation, and therefore pattern as proper names, along with definite predicate NPs. Finally, we would assume that this is an interpretation strategy that is available for all proper names in Biblical Hebrew.

A second argument for the predicative nature of proper names in Biblical Hebrew is the following. As was outlined in Section (2.3), English proper names can have authentic NP interpretations as the complements of numerical determiners or the definite article, and this is not the case in Modern Hebrew. There is evidence that personal names in Biblical Hebrew *do* behave like other common nouns, although such cases are far from frequent. There are some examples of proper names appearing with the definite article, most commonly in constructs:

(24) a. *xaci ševet*   **ha-**menaše   (Deut 3:13 et passim)
        half tribe.M.S.CS   DEF-Menashe
        'half the tribe of Menashe'
     b. *ševet*   **ha-**levi   (Deut 10:8)
        tribe.M.SG.CS   DEF-Levi
        'the tribe of Levi'

For the data in (23a) and (23b) to be interpretable, we must assume that proper names in annex position are NPs and not DPs, because if they were to be DPs, these examples should be infelicitous, which they are not. Additionally, if personal names were full determiner phrases, as they are in Modern Hebrew, they would not be able to take the definite article, as they would be closed to additional modification.

Additionally, there are rare occasions in which proper names head sentential predicates and allow restrictive relative clauses. Note that in the following examples, the English translation uses a definite determiner:

(25) a. *hu* **'*aharon u-moše*** *'ašer amar adonay la-hem*   (Ex 6:26)
        he   Aharon and-Moses   that said Lord   to-them
        'This was (the) Aharon and Moses whom the Lord told'
     b. *wa-yehi bi-yəmey   axašweyroš [...]*
        and-was in-day.M.PL.CS Ahasuerus [...]
        hu **axašweyroš** *ha-molex me-hodu   ve-ad kuš* (Esther 1:1)
        he   Ahasuerus   COMP-ruled from-India and-to Cush
        'And it was in the days of Ahasuerus [...] this is (the) Ahasuerus who ruled from India to Cush.'

While these examples are rare, they support the analysis of personal names as predicates.

These names are predicates both syntactically and semantically, but they appear to shift to a referential interpretation (although they remain NPs) when they appear in argument position. This process, through which names shift from predicate to argument interpretation, is independently necessary in order to enable bare arguments to have an argument interpretation. This can also occur with bare indefinites, as shown in (25):

(26) wa-yelek   'iš   mi-beyt   lewi   (Ex 1:1)
      and-went man.M.SG   from-house.M.SG.CS Levi
      'And a man from the house of Levi went.'

This process also occurs with definites, which are predicates denoting singular sets (Doron and Meir 2013), with uniqueness following from the singularity constraint on the set.

(27) wa-yiqqod   ha'iš   (Gen 24:26)
      and-bowed down   DEF-man
      'And the man bowed down.'

Proper names, like definites, denote singular sets, and we therefore assume that the same process applies, and therefore there is reason to believe that proper names are indeed predicates in Biblical Hebrew, with referential interpretation coming as a result of choice functions, which will be explained in the next section.

4 Syntactic and semantic analysis

We propose an analysis of Hebrew proper names in the spirit of Longobardi (1994), who analyses referential expressions as DPs, and predicate expressions as NPs. In Longobardi's (1994) analysis, proper names begin as predicate NPs. They become referential when the determiner position is appropriately filled. This position can either be filled by a determiner (as in (27a–c)) or the NP itself can raise to fill the determiner position (as in (27d)). Examples (27a–b) are from Italian (taken from Longobardi 1994: 622, 651):

(28) a. *La Callas*
b. *il Gianni*
c. *the Johns I have known*
d. *John*

Following this analysis, we assume that in Biblical Hebrew, proper names are NP predicates and denote sets. They occur as predicates in constructs, in sentential predicate position and with the definite marker as in the examples in (24).

As our analysis does not require Biblical Hebrew to have DPs, we do not need to assimilate and adapt Longobardi's (1994) theory, developed in the later parts of the 1994 paper, on how NPs raise to fill the D position, and can instead take the approach that this position may not exist at all. Instead, we follow von Heusinger (1997) and Winter (2000) and assume that referentiality can be expressed via *choice functions*.

A choice function is a function from sets (predicate interpretations) to objects. Choice function $\mathcal{f}$ maps any set X onto a chosen element $\mathcal{f}(X) \in X$. For our purposes, we think of $\mathcal{f}(X)$ as a contextually salient object, so $\mathcal{f}$ introduces referentiality. When X is a definite predicate, with a singleton denotation, there is only one choice possible for $\mathcal{f}$, which otherwise would select a member of the set at random.

We capture the difference between definite predicates and predicates that are proper names by making the standard assumption that proper names are rigid and fixed in interpretation, and therefore have the same denotation in all evaluation worlds, whereas normal definites are not.

Standardly, this is expressed by assuming that the definite *the teacher* is analysed as σ(TEACHER), which in every world denotes the person who is the unique teacher in that world, whereas the proper name *ariela* is analysed as ARIELLA, which denotes the same person (Ariella) in every world of evaluation. If *the teacher* and *ariela* are predicative NPs (as they would be in Biblical Hebrew), then their interpretations are {σ(TEACHER)} and {ARIELLA}, with the same assumptions.

The referential interpretation is created by applying the choice function $\mathcal{f}$ in such a way that $\mathcal{f}(\{\sigma(\text{TEACHER})\})$ will pick out in every world the object that is the teacher in that world. $\mathcal{f}(\{\text{ARIELLA}\})$ will pick out in every world *ariela*.

For Modern Hebrew, there is evidence that proper names enter the language at the referential type, rather than the predicate type, and we assume that proper names in Modern Hebrew are DPs, possibly as a result of historic raising of N to D, as proposed in Longobardi (1994). Here, we can follow the standard analysis and assume that *ha-mora* and *ariela* denote σ(TEACHER) and ARIELLA, with the same conditions concerning rigidity of interpretation, and we derive the same referential interpretation without using choice functions.

5 Potential implications

In support of the claim that proper names in Biblical Hebrew are not DPs, we note the absence of any explicit evidence for determiners or a grammatical determiner position in Biblical Hebrew. Standard syntactic and semantic candidates for being determiners either do not exist in Biblical Hebrew or belong to some other category. As has already been noted, in both Biblical and Modern Hebrew there is no indefinite determiner, and definiteness is marked by a feature on the noun that induces agreement on adjectives and is not a determiner. Doron and Meir (2016) suggest that that which is often analysed as the definite clitic is in fact part of state morphology. As in some forms of Aramaic such as Old Aramaic and Syriac (König 1901), there is a three-way morphological contrast in noun morphology, which is semantically interpreted.

(29) a. *bayit* 'a house'   - **absolute state** (indefinite sortal)
      b. *beyt* 'house'+annex   - **construct state** (relational)
      c. *ha-bayit* 'the house'   - **emphatic state** (definite sortal)

According to this theory, the definite *ha*-morpheme is not even a clitic, but is part of the bound morphology of the emphatic state.

In addition to the definite marker, it also appears that the demonstrative in Biblical Hebrew is not a determiner. Demonstratives are adjectival, and like adjectives, appear post-nominally, agreeing in definiteness with the noun which they modify (see (30)):

(30)   *ha-daḇar*   *ha-ze*   (2Sam 11:25)
       DEF-matter   DEF-this
       'this matter'

Quantity expressions, which in many languages are analysed as determiners, are able to act as post-nominal modifiers in Biblical Hebrew, agreeing in gender and number with the noun they modify, and therefore they too are clearly adjectival (see (31)):

(31) a. *yam-im   rabb-im*   (Gen 37:34)
         day-M.PL   many-M.PL
         'many days' (i.e. a long time)
      b. *nə'ar-ot   rabb-ot*   (Esther 2:8)
         girl-F.PL   many-F.PL
         'many girls'

Quantity can also be expressed by quantifiers, which appear in the head of constructs as in (32):

(32) *kol   ha-'anaš-im*   (Num 14:22)
      all.CS   DEF-man-M.PL
      'all the men/every man'

It is plausible to assume that these quantifiers in Biblical Hebrew are nouns, as for one thing, they occur as bare nouns, as in (33):

(33) *way-hwh   berak   'eṭ-'aḇrahām   **b-a-kol***   (Gen 24:1)
      and-God   blessed DOM-Abraham   with-DEF-all
      'And God blessed Abraham with all'

If it is the case that there are no determiners, then it is reasonable to assume that there is no D position, and hence no DP in Biblical Hebrew. If so, then proper names must be NPs in Biblical Hebrew.

As we have argued above, in Modern Hebrew, in contrast, there is evidence that proper names enter the language as the referential type rather than the predicate type, suggesting that Modern Hebrew *does* have DP structures and that proper names in Modern Hebrew are DPs. Doron and Meir (2016) argue that there is independent evidence that in Modern Hebrew the definiteness marker *ha-* is evolving from an affix expressing the definiteness feature of the noun to have full determiner function.

We have noted that in both Biblical and Modern Hebrew, place names have properties of predicate NPs, and therefore there is good reason to expect that the change in category of personal names from NP to DP does not affect place names, which remain as NP predicates in both Biblical and Modern Hebrew.

6 Conclusion

In this paper, we explored the properties of personal names in Modern and Biblical Hebrew, primarily through the frame of the construct phrase and their relative distribution. We showed that there are clear differences in distribution, and we suggested that it is not as change in the construct stage itself in the development from Biblical to Modern Hebrew that is responsible for the differences, but rather a change in the categorical status of proper names, from NP predicates in Biblical Hebrew to referential DPs in Modern Hebrew.

For Modern Hebrew, we suggest that personal names are generated as referential DPs, with the standard notion of rigidity. For Biblical Hebrew, we noted that there is no evidence for a DP position, and a great deal of evidence that proper names are NPs. We adapted the analysis of referentiality to this case, i.e. the case where there is referentiality, but no DP. We adapted Longobardi's (1994) analysis, suggesting that when names stand alone, they remain as NPs, but act as arguments rather than as predicates. Proper names and definite NPs are interpreted relative to worlds of evaluation as singleton sets, with the interpretation of proper names being rigid. The referentiality comes in via a choice function that picks out the single element of the singleton set.

This analysis predicts the differences found in distribution of personal names in construct states. The annex of the construct state is both in Biblical Hebrew and in Modern Hebrew an NP position. In Modern Hebrew this means that personal names cannot occur there, while in Biblical Hebrew they *can*, because they are NPs.

We examined additional evidence for this analysis, including the lexical composition of personal names, their taking of the definite article and ability to act as sentential predicates. We note that there are a number of exceptions to the rule, primarily place names, which will require further analysis. Further analysis is also necessary to definitively rule out the option that it is the status of the construct that has shifted between Biblical and Modern Hebrew.

Acknowledgements

This paper was presented as a joint work in March 2019, and with Susan's untimely death in July 2019 the task of writing it up fell to me. I have tried to remain faithful to the original presentation and have avoided adding anything that was not agreed upon. My thanks go to those who helped me reach this goal, particularly Keren Khrizman, Gabi Danon and Fred Landman, as well as to three anonymous reviewers. This research was funded by ISF Grant 205147 to Professor Susan Rothstein.

Abbreviations

3	third person
ABS	absolute
AUX	auxiliary verb
COMP	complementizer
CS	construct state
DEF	definite article
DOM	direct object marker
DP	determiner phrase
F	feminine
GEN	genitive
M	masculine
NOM	nominative
NP	noun phrase
PL	plural
POSS	possessive
REL	relativizer
SG	singular

References

Borer, Hagit. 2009. Afro-Asiatic, Semitic: Hebrew. In Rochelle Lieber & Pavol Štekauer (eds.) *The Oxford handbook of compounding*, 491–511. Oxford: Oxford University Press.

Danon, Gabi. 2017. Imagine no possession: *John Lennon* in the construct state. In Noa Brandel (ed.), *Proceedings of Israel Association for Theoretical Linguistics 2017* (MIT Working Papers in Linguistics 89), 49–68. https://www.iatl.org.il/?page_id=1597 (checked 04/08/2019).

Doron, Edit & Irit Meir. 2013. Amount definites. *Recherches Linguistiques de Vincennes* 42. 139–165.

Doron, Edit & Irit Meir. 2016. The impact of contact languages on the degrammaticalization of the Hebrew definite article. In Edit Doron (ed.) *Language contact and the development of Modern Hebrew* (Studies in Semitic Languages and Linguistics 84), 281–297. Leiden: Brill.

GKC = Kautzsch, E. 1910. *Gesenius' Hebrew grammar*. Second English edition as edited and enlarged by the late E. Kautzsch, revised in accordance with the twenty-eighth German edition (1909) by A. E. Cowley. Oxford: Clarendon Press.

Hess, Richard S. 2013. Names of people: Biblical Hebrew. In Geoffrey Khan (ed), *Encyclopedia of Hebrew language and linguistics*. https://referenceworks.brillonline.com/entries/encyclopedia-of-hebrew-language-and-linguistics/names-of-people-biblical-hebrew-EHLL_COM_00000229 (checked 16/11/2019).

von Heusinger, Klaus. 1997. Definite descriptions and choice functions. In Seiki Akama (ed.), *Logic, language and computation* (Applied Logic Series 5), 61–91. Dordrecht: Kluwer.

König, Eduard. 1901. The emphatic state in Aramaic. *The American Journal of Semitic Languages and Literatures* 17(4). 202–221.

Kremers, Joost. 2003. *The Arabic noun phrase: A minimalist approach*. Utrecht: LOT.

Longobardi, Guiseppe. 1994. Reference and proper names. *Linguistic Inquiry* 25(4). 609–665.

Rothstein, Susan. 2012. Reconsidering the construct state in Modern Hebrew. *Rivista di Linguistica* 24(2). 227–266.

Rothstein, Susan. 2017. Proper names in Modern Hebrew construct phrases. In Tanja Ackermann & Barbara Schlücker (eds.), *The morphosyntax of proper names*. [Special issue]. *Folia Linguistica* 51(2). 419–451.

Schniedewind, William M. 2013. *A social history of Hebrew: Its origins through the Rabbinic period*. New Haven: Yale University Press.

Winter, Yoad. 2000. What makes choice natural? In Urs Egli & Klaus von Heusinger (eds.), *Reference and anaphoric relations* (Studies in Linguistics and Philosophy 72), 229–245. Dordrecht: Kluwer.

Damaris Nübling
Von Heidel- nach Bamberg, von Eng- nach Irland? 'From Heidel- to Bamberg, from Eng- to Ireland?'

On the delimitation of appellative proper names and genuine proper names*

Abstract: This article focusses on German proper name compounds, whose head designates the category membership of its referent ("appellative proper names"), as in *Feldberg*, where *-berg* refers to a mountain (*Berg* 'mountain'). Furthermore, German has names that look very similar and yet, they are fully established names ("genuine proper names") as they do not have any semantic relation to their referent (*Heidelberg* as a city name). However, semantic (mis)match with the referent is not the decisive criterion in distinguishing the two name categories. Rather, it is their grammatical behaviour, i.e. loss of headedness, referential instead of lexical gender assignment, specific genitive and plural inflection, diverging behaviour of determiners and the increasing impossibility of partial ellipsis.

Keywords: appellative proper name, common noun, ellipsis, genuine proper name, proprialisation.

1 On determining appellative proper names

Within the class of proper names there is a hybrid type, which bares remnants of appellative function. Thus, appellative proper names (sometimes also "common-based proper names" or "category-designating proper names") do not consist of frozen or semantically opaque lexical material, as is the case e.g. in surnames such as *Schäfer* lit. 'shepherd'. Rather, these names consist of appellative material that still unfolds characterizing function: Appellative proper names usually have a part that can be connected to a contemporary common

---

Damaris Nübling: Johannes Gutenberg-University Mainz. E-mail: nuebling@uni-mainz.de.

* This article is a shortened version and translation of Nübling (2018). I am very grateful to Anke Lensch for the translation.

https://doi.org/10.1515/9783110672626-007

noun (appellative) which is usually a free lexical item. Members of this group of appellative proper names contain parts (underlined in the following examples) that characterize the denoted object, consider *Bodensee* lit. 'Boden-lake / Lake Constance'; *Eiffelturm* 'Eiffel Tower', *Schwarzwald* 'Black Forest'; *Johannes Gutenberg-Universität* 'Johannes Gutenberg-University'; *Wilhelmstraße* 'Wilhelm Street'; *Alexanderplatz* 'Alexander Square'. Thus, these complex appellative proper name compounds feature onymic parts serving an identifying purpose as well as characterizing appellative function (Fritzinger 2018; Harweg 1998: 310–316).

Harweg (1983; 1997) was the first to explore this type of so-called appellative proper names in German and defines them as follows:

> Die Eigennamen unterscheiden sich [...] darin, ob sie die Kategorie, der ihre Träger zugehören, mitbezeichnen [...]. Diejenigen, die sie mitbezeichnen, nenne ich Gattungseigennamen, diejenigen, die es nicht tun, reine Eigennamen. [...] Der Gesamtausdruck [des Gattungseigennamens – DN] ist also in erster Linie Eigenname [...] (Harweg 1983: 159–160).

> Proper names differ in whether they denote the category to which their referents belong [...]. Those proper names that contain a part naming their category are called appellative proper names. Those names that do not contain such an element are called genuine proper names [...] Thus, the whole expression [of the appellative proper name – DN] is first and foremost a proper name [...] (Harweg 1983: 159–160).

The appellative elements cannot be omitted. While in the examples above, the second element is the appellative part of the proper names, in the following examples, it is the first element that has this function: *Villa Hügel*, *Palais Schaumburg*. Omission of the appellative elements of these names is not permissible as the resulting structure would no longer be associated with its actual referent.

In this contribution, I exclusively analyse the name type that contains an appellative last part. I follow Harweg's terminology and stress what he points out in the last sentence: Despite the hybrid characteristics of appellative proper names, they are a sub-type of proper names. No one would challenge that *Hegelstraße* 'Hegel Street' or *Bismarkstraße* 'Bismark Street' are street names and that *Johannes Gutenberg-Universität* 'Johannes Gutenberg-University' or *Albert-Ludwigs-Universität* 'Albert-Ludwigs-University' are names of universities. The most important criterion is that these names are mono-referential.

Distinguishing appellative proper names from genuine or (following Fritzinger 2018) prototypical proper names is not trivial. Does *Schwarzwald* 'Black

Forest' still retain the semantics of its last element or is it rather the name of a mountain range? Is *Deutschland* 'Germany' (*-land* 'country', lit. 'land of the Germans') an appellative proper name, while *Dänemark* 'Denmark' (*-mark* 'march') or *Frankreich* 'France' (*-reich* 'empire') are not? Does there conversely have to be a mismatch between the literal denotation of the term and the referent so that the respective term qualifies as a genuine (or prototypical) proper name?

In all probability, most proper name compounds have passed through a stage where they were members of the appellative proper name subclass. It is generally known that (almost) all terms that are now proper names have evolved out of terms that used to have a direct literal semantic link to the place or city they denote today.[1] A lot of genuine proper names – mostly toponyms in the wider sense – are (or used to be) morphologically complex, often they are (or used to be) compounds. The loss of literal meaning and the freezing of the compound which leads to the formation of a genuine proper name is not abrupt. Moreover, it can affect one part of the compound earlier than another part. Hence, proper names with appellative elements are probably "younger" proper names. The diachronic trajectory of change can be described as follows:
1) Compound ending in a common noun (*Weinberg* lit. 'wine mountain/vineyard') >
2) Proper name with a part that has an appellative function (*Feldberg* 'Feldberg mountain', *-berg* 'mountain')[2] >
3) Genuine proper name (*Heidelberg* 'city of Heidelberg')

There is a very fine line distinguishing (3) proper name compounds (*Heidelberg* as a city) from (2) proper names with parts that have appellative function (*Feldberg* as a mountain). More often than not, the distinction is very subtle and thus difficult. The historically semantically transparent part does not need to undergo formal changes in the process of becoming semantically opaque. Only once it is semantically opaque, the whole name can qualify as a genuine proper name.

---

1 Here, always complex proper names of the type *Schwarzwald* 'Black Forest', *Feldberg* 'Feldberg mountain', *Bodensee* 'Lake Constance' are meant. It is obvious that this cannot apply for simplex names such as *Köln* 'Cologne' or *Berlin* 'Berlin'.

2 The case of *Feldberg* is actually more complicated. On the one hand, *Feldberg* refers to two different mountains. One is situated in the Black Forest, the other in the Taunus mountains. Hence it has to be considered an appellative proper name. On the other hand, *Feldberg* furthermore denotes a small city in Baden-Württemberg, which means that in this case *Feldberg* has genuine proper name status and belongs to Type 3.

Thus, it is not possible to discern the exact status of the respective component's semantic dissociation by exclusively assessing its formal representation. Due to formal equivalence of the appellative and onymic part, it is impossible to determine whether there is already an increase in semantic opacity of an element in question. Semantic dissociation with the descriptive meaning of the formally equivalent common noun is a sign of a shift towards proper name status.

It is indisputable that semantic mismatch is indicative of genuine proper name status: There are city names ending in *-berg* lit. '-mountain' (*Heidelberg*); *-bach* lit. '-stream' (*Ansbach*), *-hafen* lit. '-harbour' (*Ludwigshafen*), *-see* lit. '-lake' (*Falkensee*), *-land* '-country' (*Westerland*), *-feld* lit. '-field' (*Bielefeld*), *-wald* lit. '-forest' (*Greifswald*). Here, the literal meaning is not activated. Yet, this paper argues that semantic incongruence cannot be regarded as the only criterion for determining whether or not a particular term in question has reached full proper name status.

While Harweg (1983: 164, 167) considers *England* and *Deutschland* as appellative proper names (due to *-land*), in this paper I argue that although both contain bases that have a semantically transparent referent, they – due to their grammatical behaviour – have already become genuine proper names (denoting states, just like *Ungarn* 'Hungary' or *Schweden* 'Sweden'). Similarly, I postulate that *Neustadt, Bierstadt, Darmstadt*[3] are genuine proper names (denoting cities), just like *Heidelberg* and *Berlin*. This means that at some point, appellative proper names acquire genuine proper name status. This process is not necessarily accompanied by an evolving semantic mismatch. Ultimately, grammatical properties can reveal much more about the exact status of the respective name.

From a diachronic point of view, the transgression from appellative proper name to genuine proper name is a continuum. Although the main aim is to draw a line between the two categories, it is expected to find an abundance of boundary-straddling types.

---

3 Contemporarily, *Neustadt* could be literally translated as 'new city', *Bierstadt* as 'beer city' and *Darmstadt* as 'intestine city'. Yet, there is a semantic mismatch when combining the literal meaning of the two components and comparing it to their reference as city names.

2 Criteria for the distinction of appellative proper names and genuine proper names

In the following sections, I argue that it is possible to grasp the transition from appellative proper name to genuine proper name by employing grammatical rather than semantic criteria. I offer the most important delineating criteria, however in differing levels of detail.

2.1 Semantic (mis-)matches

It has previously been established that a semantic mismatch is a property of genuine proper names: *Heidelberg* does not feature any component overtly signifying it refers to a settlement. The city name *Düsseldorf*, lit. '*Düssel*-village', at least refers to a settlement, which is however a big city today and not a small village anymore. In this case, the referent has changed: It grew from the size of a village to the size of a city. Its name however remained the same. *Schwarzwald* 'Black Forest' and *Odenwald* 'Forest of Odes' raise similar questions: Do they refer to a mountain range rather than a forest? To a native speaker, an utterance such as *the highest mountain of the Black Forest has a summit of almost 1500 m* appears less marked than *the Black Forest is a conifer forest*. The literal "outgrowing" and the ensuing increasing semantic distance between the category class and the respective referent indicates that an appellative proper name has become a genuine proper name.

Often, only the descriptive meaning of the respective appellative has undergone semantic change. Thus, Middle High German *burc* not only denoted 'castle', but also 'town, city'. In modern German, *Burg* only denotes a 'castle'. Yet, *Freiburg* and many other city names still contain the *-burg* element and yet, there is not necessarily an actual castle present, or at least its presence has become irrelevant. The descriptive meaning of the individual parts has long been disassociated from the name and yet, the form of the name has been preserved. Accordingly, the semantic mismatch is not rooted in changes affecting the object that is named but in changes of the descriptive meaning of the particular linguistic element. The pathways resulting in a semantic mismatch are of no concern for the proper name status.

With those names that still involve a partial semantic match (as in *Deutschland* 'German-<u>land</u>', *Neu<u>stadt</u>*, 'New-<u>city</u>', *Darm<u>stadt</u>*, 'Darm-<u>city</u>'), the semantic criterion is not decisive, it is rather their grammatical behaviour. As form mostly follows function, it is likely that the semantic connection *Deutsch-land* 'land, state' has long been deactivated, as is obviously the case with *Frank-reich* 'empire of the Franks' and *Luxem-burg* '-castle'. In addition to Harweg's criteria, there are some other criteria that underpin the argument that these have genuine proper names status. The dissociation from the common noun becomes evident subtly and morphologically and thus it is closely connected to the question whether the respective elements morphologically and grammatically behave like a common noun (appellative) or rather like a (proper) name.

2.2 Loss of headedness and increase in referential gender assignment

As is demonstrated in Fahlbusch and Nübling (2014), in the process of becoming a proper name, the gender assignment of a name shifts from the original, inherited appellative gender of the common noun, thus from the so-called morpho-lexical gender to a new and different, the so-called referential gender. Referential gender is determined by the object class of the referent of the proper name. Some examples can illustrate this: Names of ships always have feminine gender. Other names (*Kaiser Wilhelm* (m.), *Europa* (n.)) or common nouns (*Albatros* (m.) 'albatross') that are used to name ships, undergo category change. As soon as they are used as ship names, their original gender is ousted by referential gender: *(der) Kaiser Wilhelm* (m.) > *die Kaiser Wilhelm* (f.), *der Albatros* (m.) > *die Albatros* (f.), *Europa* (n.) > *die Europa* (f.).

The far younger name class of magazines is currently establishing and strengthening feminine gender as its referential gender. Thus, feminine gender increasingly ousts the original gender: *die BILD, die Capital*, occasionally *die Playboy*, however still *der SPIEGEL, der Focus*. Some titles that were formerly inflected in the plural, such as *die BNF* (< *Beiträge zur Namenforschung*), shift to singular feminine gender (which can be seen at the modifier: *die neuen* (pl.) *BNF* > *die neue* (sg.) *BNF*). Notably, the formal syncretism of plural and feminine forms (*die, sie, ihre*) pave the way for this process. Car names are masculine: *der Mercedes* (*Mercedes* is a Spanish first name for females), *der Polo, der Fiesta, der Corsa*. This process shows to be productive concerning acronyms (*der TT*). In contrast, names of restaurants have neutral gender (*das* (n.) *Steigenberger; das* (n.) *Turm* < *der* (m.) *Turm* 'the (m.) tower'). Thus the lexical gender of the original lexeme is replaced by the referential gender.

In those cases where the original lexeme is a compound, in the process of the shift from appellative proper name to genuine proper name, the Right-hand Head Rule is overruled. The right-hand element (the head) loses its gender-assigning function. In contrast, in names that involve appellative elements, these elements retain their gender-assigning function, their referential righthand element maintains its head-designating status. Only genuine proper names no longer have a head. Even if they are complex, there is no hierarchy anymore among their components.

The gender criterion demonstrates that *Neustadt, Darmstadt* etc. are genuine proper names: *Stadt* 'city' as a descriptive common noun is feminine, yet without exception, German city names are assigned neuter gender, thus *die* (f.) (*Groß-, Provinz-, Haupt-*) *Stadt* 'city, provincial city, capital' (all are common nouns) against *das* (n.) *schöne* ('beautiful') *Neustadt, Darmstadt, Ludwigsstadt,* etc. (all are names). Consequently, any city name that ends in *-stadt* '-city' is a genuine proper name. The neuter gender of city names is assigned referentially. Thus, neither the sum of the meaning of the city name's parts, the complexity of the respective city name nor any transparent lexical remnants in the city name have an influence on the categorical gender of city names. City names constitute their own name class and this category has long been fully established. To date the exact point in time marking the emergence of the city name class has not been ascertained.

Steche (1927: 81) writes with regard to city (and state) names: "Aber diese Einheitlichkeit [neutrales Genus] ist geschichtlich sehr jung; im Althochdeutschen und Mittelhochdeutschen konnten die Namen der Orte und Länder jedes Sprachgeschlecht haben" ['However this unity [neuter gender] is historically very young; in Old High German and Middle High German names of cities and states could have any gender']. Paul (1917/1968) observes that in former times foreign city names frequently had feminine gender, as their source language's grammatical gender was adopted. Often the source languages were Latin and Greek: Lat. *Roma* (f.) > *Rom* (f.), *Colonia* (f.) > *Köln* (f.), just as *Rhodos* (f.), *Carthago* (f.), *Athen* (f.). Today, all of these cities have neutral gender in German. Concerning German city names, he writes that they are usually compounds ending in e.g. *-burg* '-borough', *-dorf* '-thorp', *-heim* '-ham, home', *-stadt* '-stede, city', *-weiler* 'village, hamlet'. He continues that among these, the grammatical gender of the second component must have been decisive, but in his day, all city names had already been assigned neuter gender. Paul assumes that the grammatical gender of the second part of the city name determined its gender (cf. Paul 1917/1968 §117). Paul further hypothesized that the neuter gender of the whole class was fed by the most frequently occurring second elements such as *-dorf* '-thorp, village', *-heim* '-ham, home', *-tal* 'dale' with lexical neuter gender.

This means that whenever there is a mismatch of the grammatical gender of a name and the original grammatical gender of the second element of this name, this name must be a genuine proper name. However it is also possible that the referential gender of a proper name class and the original lexical gender of the last element are the same. Structurally, gender assignment has long become independent from the head, as assumed in the case of *Deutschland* 'German-land, Germany' (n.). As there are only three grammatical genders, the likelihood of overlap is not low. The common noun *Land* (n.) 'country' has neuter gender. All state names (besides very few exceptions such as *die Schweiz* (f.) 'Switzerland', *die Türkei* (f.) 'Turkey') are assigned referential neuter gender.

As state names are usually not accompanied by a determiner (which would overtly convey information about the gender), the particular state's gender only surfaces when syntax requires a determiner, i.e. when the proper name is modified by an adjective: *Das* (n.) *schöne Polen* ('the nice Poland'), *das* (n.) *schöne Luxemburg* ('the nice Luxemburg'), *das* (n.) *schöne Deutschland* ('the nice Germany').

Thus, in the case of *Deutschland* 'German-land, Germany', neither the semantics nor the state name's gender clearly reveals its genuine proper name status. Yet, characteristic zero-determiner uses (see 2.3), the fact that the name is resistant to ellipsis in that none of its parts can be omitted (see 2.4), and the way *Deutschland* undergoes inflection (see 2.5) show that it has reached the status of a genuine proper name. In most cases, however, semantic mismatches are in line with the formal developments.

2.3 Diverging behaviour of determiners

Appellatives constitute a prototypical noun class as they are comprised of countable and contoured concrete nouns, which are obligatorily accompanied by different determiners. Most commonly, these determiners are definite or indefinite: *die* (f., definite) *Stadt* 'the city', *eine* (f., indefinite) *Stadt*, *manche* (f., pronoun) *Stadt* 'some city'. The determining function of these determiners is pivotal. As they are in opposition with each other, they constitute a genuine paradigm. This maximal requirement is not met by appellative proper names anymore.

Appellative proper names of our type (with appellative last part) are often accompanied by a definite determiner (which is in congruence with the last part of the name). However, the respective determiner's paradigm is not used in its entirety (except under highly specific circumstances): *der* (m.) *Bodensee* ('the Lake Constance') in contrast to **ein Bodensee* ('a Lake Constance'), **manch Bodensee* ('some Lake Constance') vs. the common noun *der* (m.) *Baggersee* 'the quarry pond', *ein Baggersee* 'a quarry pond', *manch Baggersee* 'some quarry pond'.

One of the properties of some genuine proper name classes is that their members are always accompanied by a determiner, which is grammatically not required as definiteness is an inherent property of proper names. However, what appears to be a determiner is analysed as a classifier in Nübling (2020). Most proper name classes are accompanied by an invariant classifier. These classifiers have the appearance of determiners, yet they have already left the determiner's paradigm and they no longer have a functional link to other determiners. Members of this type are names of deserts (*die* (f.) *Namib* 'the Namib', *die* (f.) *Sahara* 'the Sahara'), mountains (*der* (m.) *Schauinsland* 'the Schauinsland', lit. 'look-into-the-country', *der* (m.) *K2* 'the K2'), cars (*der* (m.) *Mercedes* 'the Mercedes', *der* (m.) *Fiesta* 'the Fiesta') etc. (for more detail see Fahlbusch and Nübling 2014; Nübling 2020). Some proper name classes, in particular older name classes which are more established, are attested occurring without determiners: City names (Ø *München* 'Munich', Ø *New York*), state names (Ø *Polen* 'Poland', Ø *Dänemark* 'Denmark'), names of continents (Ø *Asien* 'Asia', Ø *Europa* 'Europe') and, in Standard New High German, personal names (Ø *Angela Merkel*).

Consequently, names that are not obligatorily accompanied by a determiner, or, to be more precise, a classifier, are always genuine proper names. As *Deutschland* 'Germany' belongs to this group, it cannot be considered an appellative proper name. Paul (1917/1968) points out that in earlier periods, there were uses of *das* (n.) *Deutschland*. However, he observes that already in his day, names ending in *-land* 'country' and *-reich* 'empire' had completely adopted the characteristics of proper names (Paul 1917/1968, Vol. 2, Part III: 162).[4]

Today, the standard use Ø *Deutschland* rules out appellative proper name membership. Similarly, *Luxemburg* 'Luxembourg' is not an appellative proper name class member and there are three reasons leading to this assumption: *Luxemburg* is a state and not a *Burg* (f.) 'castle', it has neuter gender and not feminine gender and it is used without a determiner. Thus, the presence of determiners does not indicate proper name status, in turn occurrences without a determiner do indicate proper name status (contradicting Harweg 1983: 166–167). The so-called syntactic (or secondary) determiner, which is obligatory

---

4 "Früher sagte man auch *das Deutschland*. Jetzt haben die Bezeichnungen mit *-land* wie die mit *-reich* ganz den Charakter von Eigennamen angenommen, so auch *Dänemark*, nachdem es zur Bezeichnung eines Staats geworden ist [...]" (Vol. 2, Part III: 162). 'In former times, it was common to say *das Deutschland* 'the Germany'. Contemporarily, the denominations with *-land* 'state' and *-reich* 'empire' have fully adopted proper name status. This also applies to *Denmark* after it became the name for a state [...]'.

whenever there is a modifier, has already been mentioned. It has a different status (for a more detailed discussion of determiners combining with proper names see Fritzinger 2014, 2018).

2.4 The (im-)possibility of partial ellipsis

To date, studies in onomastics have mostly not considered whether it is still possible to separate or coordinate parts of complex proper names. Thus, there has hardly been any research to answer the question whether it is possible to leave out a part of a proper name if it occurs in coordination with another proper name sharing the same final part (Fritzinger 2018). According to Kempf (2008), contemporarily, ellipsis of the second element (as in *Juden- oder Christentum* 'Jewish religion or Christian religion', *-tum* is a cognate of English *-dom*) is far more frequently attested than ellipsis of the first element (as in *Pferdezucht und -haltung* 'horse breeding and horse rearing'). According to Bergmann (2018: 23–26), this phenomenon is called "coordination reduction".

In the context of this paper I focus on the first kind of ellipsis, where the last element can be left out, as the appellative part of names is usually the last part of the respective name: *von der Hegel- in die Bismarckstraße*, 'from Hegel street to Bismarck street', *vom Helmholtz- zum Goethe-Gymnasium* 'from Helmholtz High School to Goethe High School'. Since the first and the last element of complex names constitute independent phonological words, there is no phonological rule that would disallow this kind of ellipsis (for more detail see Kempf 2008).

In German, word ellipsis has undergone remarkable diachronic change (cf. Kempf 2010). Up until the 15th century, inflectional endings could be omitted (in this case the plural inflection of *underrichtungen* 'lessons': *mit disen und andern underrichtung[en] und leren* lit. 'with these and other lesson[s] and teachings' - 'with these and other lessons and teachings'). From the 16th century onwards, derivational morphology and compounds is attested in elliptic structures: *in erober-Ø vnd plünderung der Statt*: *-ung* '-ing' is omitted 'during the conquest and ransacking of the city'; *das Richter-Ø oder Beysitzerambt*: *-amt* 'office' is omitted 'the office of judge and the office of assessor'. During the construction's heyday in the 17th and 18th century, the omission of parts of proper names was possible: *Lief- und Russland* lit. 'Liv- and Russianland' – 'Livonia and Russia', *in Holl- und Engelländischen* 'in Holl- and England' – 'in Holland and England'. Kempf (2010: 361) finds that these structures are no longer grammatical as, according to her, morphological segmentation of proper names is impossible: **Sie spielte in verschiedenen Vereinen in Deutsch- und England* 'She played for several clubs in Germany and England', lit. 'Germanland and England' (Kempf

2010: 361). Whether Kempf's use of the asterisk should be applied categorically and systematically will be put to the test in the following corpus analysis.

I hypothesize that genuine proper names cannot be coordinated in elliptic structures, however I expect that appellative proper names can be coordinated in this way. What is more, I assume that the loss of the possibility of a particular name to occur in coordinating structures involving ellipsis is part of a diachronic freezing or fixation process.

Table 1: Word ellipsis delineating appellative proper names from genuine proper names.

1) Common nouns	2) Appellative proper names	3) Genuine proper names
ihr Heimat- und Vaterland 'her home~~land~~ and (her) fatherland' *zwischen Groß- und Kleinstadt* 'between great ~~city~~ and small city' – 'between city and small city'	*zwischen Hegel- und Bismarckstraße* 'between Hegel~~street~~ and Bismarckstreet' *zwischen Boden- und Titisee* 'between ~~Lake~~ Constance and Lake Titisee'	**zwischen Darm- und Bierstadt* 'between Darm- and Bierstadt' **von Heidel- nach Bamberg* 'from Heidel- to Bamberg' *?/*zwischen Eng- und Irland* 'between Eng- and Ireland'
Ellipsis of a part of the compound is possible		Ellipsis of a part of the name is no longer possible

Even though some genuine proper names, such as *Darmstadt* (lit. 'intestine city') (3) appear to be combinations of two free lexemes, none of the parts can be omitted and thus they cannot be found in elliptic structures. Probably this applies to all proper names referring to cities (**von Heidel- to Bamberg* 'from Heidel-Ø to Bamberg'), even in those cases where the last component of the city name is *-stadt* '-city'. The sequence ?/* in (3) indicates that the increase in invariability is a slow process.

Section 2.4.2 demonstrates that corpora contain more relics than would be expected. Personal names and place names show some similarities in that both cannot be part of elliptic structures. Thus, it is not possible to omit parts of personal names and hence they are not found in elliptic structures (consider **die Abgeordneten Brink- und Ackermann* 'the MPs Brink~~mann~~ and Ackermann'; in analogy: **the MPs Cole-Ø and Kimmelman*).[5] In contrast, appellative proper names do occur in elliptic structures as in (2): *zwischen He-*

---

[5] As this appears unquestionable, I did not test this in a corpus analysis. Nevertheless, this assumption still awaits empirical verification.

gel- und Bismarckstraße 'between Hegel~~street~~ and Bismarckstreet', *zwischen Boden- and Titisee* 'between ~~Lake~~ Constance and Lake Titisee'.

This raises the interesting question of whether or not common noun compounds and appellative proper names can be coordinated in elliptic structures: *?sie fuhr von der Hegel- in die Vorfahrtsstraße* 'she drove from Hegel~~street~~ into the main street'; *?sie fuhr von der Vorfahrts- in die Hegelstraße* 'she drove from the main ~~street~~ into Hegelstreet'. Whether or not there are indeed hybrids of (1) and (2) in Table 1 was ascertained in a corpus analysis (DeReKo 2017-II).[6] A corpus search of the type "X- und Ystraße" 'X- and Ystreet' showed that in those cases where "Ystreet", the second element, was not a proper name (e.g. *Parallelstraße* 'parallel street', *Einkaufsstraße* 'shopping street'), its counterpart "X-" is also a common noun compound (e.g. *Quer- und Parallelstraße* 'intersecting and parallel street').

Thus, purely appellative compounds (Type 1) are hardly attested in coordination with appellative proper names (Type 2), although they share the same second element (i.e. the type **von der Vorfahrts- in die Hegelstraße* 'from the main ~~street~~ [common noun] into Hegelstreet [appellative proper name]' hardly ever occurs).

With regard to ellipsis, appellative proper names can clearly be delineated from genuine proper names (Type 3).

2.4.1 *Schwarz-, Oden- und Pfälzerwald* 'Black Forest, Forest of Odes and Palatine Forest' in elliptic structures

Now we turn to the transition from appellative proper name (Type 2 in Table 1) to genuine proper name (Type 3). Genuine proper names can no longer be coordinated in elliptic structures. As the shift towards genuine proper name is gradual and continuous, it is likely to face variation. Thus, it is expected to find coordinated structures where ellipsis may still be possible but is rarely attested. Furthermore, it is expected to find particular names that have become inseparable or almost inseparable, or names that can only occur in particular combinations. Moreover, it should be determined whether the sequence of the elements in coordination is variable (e.g. Is *Oden- und Schwarzwald* attested as often as *Schwarz- und Odenwald*?). All of these questions remain unanswered, require a corpus-based analysis and cannot be fully accounted for within the scope of this paper.

---

6 I am indebted to thank Anne Rosar for conducting this search.

Yet, this paper constitutes a first attempt to provide some initial insights. The IDS-corpus *Deutsches Referenzkorpus* (DeReKo 2017) mainly contains newspaper texts. Thus, it consists of more formal, partially diachronic data (from the 19th and early 20th century). In addition, I base my analysis on the web-based DECOW-corpus, which contains contemporary internet texts (from forums, blogs and comment sections). This corpus is thus comprised of more colloquial language.

Both "Schwarzwald und/oder/zum/in den ('and/or/to/into') Xwald" (no ellipsis) as well as "Schwarz- und/oder/zum/in den Xwald" (with ellipsis) were searched in both corpora. In the same way, I searched *Odenwald* and *Oden-* as well as *Pfälzerwald / Pfälzer Wald* and *Pfälzer-* as well as *Pfälzer* (see Table 2 and 3). Despite of the immense size of the corpora (DeReKo: ca. 32 bio words, DECOW: ca. 15 bio words), the number of hits is often very small. These results should neither be overinterpreted nor should they be discounted.

Table 2 is based on DeReKo whereas Table 3 is based on DECOW 16 (see Schäfer and Bildhauer 2012; Schäfer 2015). The lowest line of Table 2 shows that with 80%, (non-elliptic) full forms dominate over 20% of elliptic structures. In DECOW (Table 3), full forms display an even stronger preponderance amounting to 88%. It appears that there are no particular concrete names (the greyed out table elements) that stand out: Consequently, it doesn't matter whether *Schwarzwald* 'Black forest', *Odenwald* 'Forest of Odes' or *Pfälzer Wald* 'Palatine Forest' serve as the first element, this appears almost irrelevant. The tables do not show what is represented by X and yet, there is not a lot of variation: In the majority of attestations, it is either *Schwarz-* or *Oden-*.

Table 2: Elliptic structures involving -*wald* '-forest' (DeReKo W-archive).

	no ellipsis		ellipsis	
Appellative proper names	Schwarzwald in den Xwald	1	Schwarz- in den Xwald	0
	Schwarzwald zum Xwald	0	Schwarz- zum Xwald	0
	Schwarzwald oder Xwald	2	Schwarz- oder Xwald	3
	Schwarzwald und Xwald	38	Schwarz- und Xwald	10
	total: Schwarzwald & Xwald	41 (76%)	total: Schwarz- & Xwald	13 (24%)
	Odenwald zum/in den Xwald	1	Oden- zum/in den Xwald	0
	Odenwald oder Xwald	11	Oden- oder Xwald	3
	Odenwald und Xwald	76	Oden- und Xwald	11
	total: Odenwald & Xwald	88 (86%)	total: Oden- & Xwald	14 (14%)
	Pfälzerwald oder Xwald	2	Pfälzer- oder Xwald	4
	Pfälzer Wald oder Xwald	3	Pfälzer oder Xwald	1
	Pfälzerwald und Xwald	43	Pfälzer- und Xwald	17
	Pfälzer Wald und Xwald	14	Pfälzer und Xwald	0
	total: Pfälzerwald & Xwald	62 (74%)	total: Pfälzer- & Xwald	22 (26%)
	all full forms:	**191 (80%)**	**all elliptic structures:**	**49 (20%)**

What is more, *und* 'and' is the most frequently attested conjunction by far and it is followed by *oder* 'or'. Prepositions are hardly attested, neither in elliptic nor in non-elliptic structures. This is confirmed by the data in both tables. There is graphematic variation between *Pfälzer Wald* (two words) and *Pfälzerwald* (solid spelling), in all the variant with solid spelling is more frequent. In elliptic structures, spellings with a hyphen dominate (in particular in Table 2). Thus, the hyphen in the *Pfälzer-* variant shows the need for coordination more noticeably than the *Pfälzer* variant without a hyphen.

Table 3, which is based on internet data, contains some additional information: In addition to limiting the search to names ending in -*wald* '-forest' the analysis of common nouns was limited to words beginning in *Stadt-* 'municipal' or *Regen-* 'rain' and/or *Xwald* 'forest'. Combinations involving *Xwald* 'forest' exclusively feature common nouns (e.g. *Stadt- und Gemeindewald* 'municipal ~~forest~~ and community forest', *Regen- und Nebelwald* 'rain ~~forest~~ and cloud forest'). Hybrids of common nouns and proper names are not attested. This corroborates the hypothesis that *Schwarz- und Odenwald* are more likely to refer to mountain ranges rather than the woods that cover these mountains.

Table 3: Elliptic structures with *-wald* 'forest' (DECOW 16).

	no ellipsis		ellipsis	
Appellative proper names	Schwarzwald in den Xwald	4	Schwarz- in den Xwald	1
	Schwarzwald zum Xwald	2	Schwarz- zum Xwald	0
	Schwarzwald oder Xwald	11	Schwarz- oder Xwald	4
	Schwarzwald und Xwald	122	Schwarz- und Xwald	23
	total: Schwarzwald & Xwald	139 (83%)	total: Schwarz- & Xwald	28 (17%)
	Odenwald zum/in den Xwald	3	Oden- zum/in den Xwald	0
	Odenwald oder Xwald	5	Oden- oder Xwald	1
	Odenwald und Xwald	104	Oden- und Xwald	8
	total: Odenwald & Xwald	112 (93%)	total: Oden- & Xwald	9 (7%)
	Pfälzerwald oder Xwald	3	Pfälzer- oder Xwald	0
	Pfälzer Wald oder Xwald	2	Pfälzer oder Xwald	0
	Pfälzerwald und Xwald	53	Pfälzer- und Xwald	4
	Pfälzer Wald und Xwald	30	Pfälzer und Xwald	4
	total: Pfälzerwald & Xwald	88 (92%)	total: Pfälzer(-) & Xwald	8 (8%)
	all full forms:	**339 (88%)**	**all elliptic structures:**	**45 (12%)**
Common nouns	Stadtwald und/oder Xwald 'municipal forest and/or X forest'	19	Stadt- und/oder Xwald 'municipal Ø and/or X forest'	31
	Regenwald und/oder Xwald 'rain forest and/or X forest'	27	Regen- und/oder Xwald 'rain Ø and/or X forest'	83
	all full forms:	**46 (29%)**	**all elliptic structures:**	**114 (71%)**

The last lines of Table 3 demonstrate that common nouns are far more frequently attested in elliptic structures (71%), whereas elliptic structures only amount to 12% of all appellative proper names. Consequently, regarding elliptic structures, names are less flexible, however they have not reached the point of invariance. Once a particular term has reached this point, it has reached genuine proper name status. Thus, the corpus data can reveal that names ending in *-wald* 'forest' are currently transitioning from appellative proper name to become a genuine proper name. The data suggests that names ending in *-wald* have all but reached the end-point of this transition process.

While *Schwarz-* 'Black-', is semantically transparent, *Oden-* 'Odes' is opaque, however there does not appear to be any effect on their sequence (see Table 4). Within their respective context, both should be recognizable as elliptic structures, compare prototypical syntagmas such as *die klare Aussicht auf Schwarz- und Odenwald* 'the clear view on the Black ~~Forest~~ and the Forest of Odes', *das Hügelland zwischen Oden- und Schwarzwald* 'the hill country between the ~~Forest of~~ Odes and the Black Forest'. The data only contain counts with a single or two digits, indicating minute differences: *Schwarzwald* seems to be placed in front of *Odenwald* a little more frequently (in both corpora) than *Odenwald* in front of *Schwarzwald*. In those cases, *Schwarzwald* is more likely to occur in elliptic structures (*Schwarzwald und Odenwald* 111 vs. *Odenwald und Schwarzwald* 100) and *Schwarzwald* is less likely to occur in elliptic structures (*Schwarz- und Odenwald* 27 vs. *Oden- und Schwarzwald* 14).

Table 4: *Oden-* before *Schwarzwald* vs. *Schwarz-* before *Odenwald*.

	Schwarz- before *Oden-*		*Oden-* before ***Schwarz-***	
Full forms	***Schwarzwald* und *Odenwald*** 'Black Forest and Forest of Odes'		*Odenwald* und *Schwarzwald* 'Forest of Odes and Black Forest'	
	DeReKo	DECOW	DeReKo	DECOW
	30	81	27	73
	111		100	
Elliptic structures	***Schwarz-* und *Odenwald*** 'Black Forest and Forest of Odes'		*Oden-* und ***Schwarzwald*** 'Forest of Odes and Black Forest'	
	DeReKo	DECOW	DeReKo	DECOW
	8	19	6	8
	27		14	

Yet, this tendency could be an effect of the general higher frequency of *Schwarzwald*: The Black Forest is more touristy, bigger in size and more famous. Thus, *Schwarzwald* is placed in front since it is the less marked component. This is common for binomials and also in more variable serialisations. Because of its fame it is possible that it more freely occurs in elliptic structures. In DECOW, *Schwarzwald* is attested 73942 times, *Odenwald* only 26688 times, amounting to only a third of the attestations of *Schwarzwald*.

2.4.2 Deutsch-, Eng- and Ireland in elliptic structures

Up to this point, names of states ending in *-land* have been assigned the status of genuine proper names. This classification was firstly based on the fact that they are not accompanied by a determiner (Ø *Deutschland* 'Germany' vs. *das Industrieland* 'the industrial state') and secondly, their last element does no longer determine the particular name's gender (*das schöne Luxemburg/Dänemark* 'the (n.) beautiful Luxemburg/Denmark'; *Burg* (f.) 'castle' and *Mark* (f.) 'march' as common nouns have feminine gender). Although the elliptic structure used in the headline appears far less acceptable than those structures involving *-wald* (compare *Schwarz- und Odenwald*), I conducted a corpus analysis (see Table 5 containing DeReKo data and Table 6 containing DECOW data).

Table 5: Elliptic structures involving -*land* 'state, country' (DeReKo W-Archive).

	no ellipsis		ellipsis	
Proper names	Deutschland bis/nach/oder Xland	1336	Deutsch- bis/nach/oder Xland	0
	Deutschland und Xland	6569	Deutsch- und Xland	9
	England bis/nach/oder Xland	729	Eng- bis/nach/oder Xland	0
	England und Xland	4422	Eng- und Xland	4
	Irland bis/nach Xland	80	Ir- bis/nach Xland	0
	Irland oder Xland	239	Ir- oder Xland	0
	Irland und Xland	2246	Ir- und Xland	3
	all full forms:	**15621**	**all elliptic structures:**	**16**

Table 6: Elliptic structures involving -*land* 'state, country' (DECOW 16).

	no ellipsis		ellipsis	
Proper names	Deutschland bis/nach/oder Xland	1963	Deutsch- bis/nach/oder Xland	0
	Deutschland und Xland	8870	Deutsch- und Xland	16
	England bis/nach Xland	493	Eng- bis/nach Xland	0
	England oder Xland	783	Eng- oder Xland	3
	England und Xland	6998	Eng- und Xland	8
	Irland bis Xland	40	Ir- bis Xland	0
	Irland nach Xland	125	Ir- nach Xland	1
	Irland oder Xland	351	Ir- oder Xland	1
	Irland und Xland	2375	Ir- und Xland	6
	all full forms:	**21998 (~ 100%)**	**all elliptic structures:**	**35**
Common nouns	Heimatland ins/oder/und Xland 'home country into/or/and X country'	45 (30%)	Heimat- ins/oder/und Xland 'home country into/or/and X country'	106 (70%)

Table 5 and Table 6 show very clear results: In both databases, in comparison to the occurrence of full forms, elliptic structures are extremely rare (although they still exist). They amount to less than 1 percent of all hits (I refrained from calculating the exact percentages in Table 5 and 6). The state of affairs is very clear: The data demonstrates that names ending in *-land* 'country' are genuine proper names. What is more, the data shows that elliptic structures hardly occur with prepositions, maybe since they denote spatial relations and therefore further describe the respective objects.

Only structures with *und* 'and' and *oder* 'or' allow for atavisms. These coordinators often correlate with uses involving abstract entities, consider *Beziehungen* 'relations' or *Unterschiede zwischen Deutsch- und Xland* 'differences between Deutsch- and Xland'.

Let's have a closer look at the more rarely attested coordination reductions:
a) "Deutsch- and Xland" 'German- and Xland': Out of 10 DeReKo-attestations in Table 5, four have their origin in a citation of Christian Morgenstern: "Und kehrte schließlich stumm nach Deutsch- und Holland um" 'and finally turned back to Germany and Holland'. "Deutsch- und Deutschland" (meaning the two separate German states that were formed after WWII, i.e. East and West Germany) are attested two times (*Grenzübergang zwischen Deutsch- und Deutschland* 'border crossing between Germany and Germany', *den kleinen Unterschied zwischen Deutsch- und Deutschland* 'the small difference between Germany and Germany'). The remainders of attestations are *in Deutsch- und Euroland* 'in German- and Euroland', *mit Deutsch- und Russland* 'with Germany and Russia' and *zwischen Deutsch- und Griechenland* 'between Germany and Greece'. In DECOW (Table 6), there are 16 attestations: 5 of these contain the Morgenstern-example, an additional hit with *Holland* is genuine as are all other hits. The second element is *Griechenland* (3x) 'Greece', *Russland* (2x) 'Russia' and once *Welschland* 'Italy and France, countries speaking Romance languages', *Engelland* (sic) 'England', *Amiland* 'informal for USA', *Ösiland* 'informal for Austria'.[7] Surprisingly, there is a genuine hybrid *Deutsch- und Umland* 'Germany and its surrounding countries': *die Tourneen durch Deutsch- und Umland werden komplett vom Wohnmobil aus erledigt* 'tours through Germany and surrounding countries are accomplished by caravan'. The genuine proper name *Deutschland* 'Germany' does not shy away from a coordination reduction with the common noun *Umland* 'surrounding countries' (*Deutsch- und Umland*). It is likely

---

7 *Amiland*, *Ösiland* and similar forms are so-called inofficial nicknames.

that the formation of this elliptic structure is fostered by the fact that both terms have neuter gender (Yet, *Deutschland* has its gender because of the referential gender of the state name class, whereas *Umland* is assigned its gender because of the grammatical gender of the common noun *Land* 'country'). Forming an elliptic structure involving *Luxem-* (n.) *und Wartburg* (f.) 'Luxemburg and Wartburg castle' (Luxemburg as a state and Wartburg as a castle) would probably be more problematic.

b) "Eng- und Xland" 'Eng- and Xland': There are four genuine attestations in DeReKo: *so verschiedenen Bürokulturen wie Eng- und Deutschland* 'as different office cultures in England and Germany', *Rockstars aus Eng- und Amiland* 'rock stars from England and the US', *die Menschen in Eng- und Irland* 'people in England and Ireland', *für Eng- und Holland die Daumen drücken* 'cross your fingers for England and Holland'. In DECOW there are eight (genuine) attestations. Two Xland attestations are *Deutschland* and *Amiland* each. There is one attestation of *Griechenland* 'Greece', *Finnland* 'Finland', *Schottland* 'Scotland' and *Niederland* (sic) 'Netherlands', each. Except for one attestation, which is used in a poem and forms a rhyme with *bekandt* (sic) 'known', the others are used in the contexts of football matches. In DECOW, there are three attestations of *Eng-* or *Xland*: *mußte man nach Eng- oder Amiland blicken* 'you had to look at Old Blighty or at the land of the Yankees for', *nach Eng- oder Irland* 'to England or Ireland', as well as another astonishing hit: *egal, ob das nun Griechen-, Eng- oder Mailand ist* 'it doesn't matter whether this is Greece, England or Milan'. Apart from the fact that this syntagma contains two ellipsis, *Mailand* 'Milan', a city name, is incorporated. *Mailand* itself is based on a folketymological analogy which led to an alteration.[8] Thus, this structure is another kind of hybrid: Two state names are coordinated with a city name. Thus two different name classes are combined.

c) "Ir- und Xland" 'Ire- and Xland': There are three attestations in DeReKo: *zwischen Ir- und Deutschland* 'between Ireland and Germany', *zwischen Ir- und Island* 'between Ireland and Iceland', *dreimal Ir- und Island* 'three times Ireland and Iceland'. In DECOW, there are six genuine attestations, three with *Schottland* 'Scotland', two with *England* 'England' and another with *Griechenland* 'Greece'. DECOW also hosts the only prepositional combination (*Freitags dampfte die Fähre mit uns an Bord von Ir- nach Schottland* 'on Fridays the ferry puffed from Ireland to Scotland with us on board') and the

---

8 Harweg (1983: 166) considers *Mailand* 'Milan' a pseudo appellative proper name.

only *oder* 'or' attestation, which features a double ellipsis: *Gründen Sie deshalb jetzt noch eine Firma in Griechen-, Ir- oder Island* 'For this reason found a company in Greece, Ireland or Iceland right now'.

The whole field of onymic coordinations requires much more comprehensive research, both in synchronic and diachronic data. As we protrude more deeply, more and new questions arise. In addition to my analysis of names ending in *-wald* 'forest' and *-land* 'country', many other second elements could be analysed (e.g. *-stadt* 'city', *-see* '-lake, sea', *-berg* 'mountain', *-gebirge* 'mountain range').

2.5 Genitive and plural inflection

In discussing inflection, we arrive at the last criterion distinguishing genuine proper names from appellative proper names. As meticulously illustrated in Nübling (2005, 2012, 2017), in German there are different inflectional paradigms for common nouns and proper names. Diachronically, this distinction is strengthened.

Roughly speaking, the inflection of proper names was originally (in Old High German and Middle High German) identical with the inflection of common nouns. Gradually, allomorphs were reduced (paradigmatic deflection) and became shorter than that of common nouns. Often, there is a tendency for zero inflection (syntagmatic deflection, see Nübling 2012; Ackermann 2018). This could be termed 'economic inflection'. Both (*-s* and *-Ø*) serve to maintain the body of the name itself as the name body is difficult to process (Zimmer 2018). Inflection with *-s* only results in a minute change of the name body, whereas *-Ø* does not result in any change.[9]

This explains why umlaut appears to be avoided e.g. in plural structures involving onymic material (consider *die Manns* 'the Mann family' vs. *die Männer* 'the men', *die Kochs* 'the Koch family' vs. *die Köche* 'the cooks'). This also applies for derivational processes, such as diminutives (*das Hannchen* 'little Hannah' vs. *das Kännchen* 'the little pot') and inflection for grammatical gender (*die Wolfin* used to refer to the wife or daughter of a man with the surname *Wolf*; *die Wölfin* denotes a female wolf; see Schmuck 2017).

---

[9] This also applies to other nouns that are difficult to process, which increases the need for them to be conserved. These nouns are usually conversions, loanwords and abbreviations (see Nowak and Nübling 2017; Zimmer 2018).

To date, there has hardly been any diachronic research to assess when proper names started not to inflect for umlaut. In contrast, umlaut remains frequent among common nouns and, even more, it is still productively applied to loanwords (*die Generäle* 'the generals', *das Skandälchen* 'the little scandal', *die Bischöfin* 'the female bishop'). For inflection of names it is assumed that there are several factors influencing how this name is inflected: If a particular name has a foreign structure (a large discrepancy to the prototypical phonological word, which in German is a trochee with a reduced vowel), if the name is less familiar and also low in frequency and if the name is obviously non-German, there is a higher likelihood for it to be preserved, i.e. remain unchanged.

Accordingly, as shown in Nübling et al. (2015: 69), the tendency of non-German toponyms for zero-inflection is the higher the more foreign (and longer) the name body is. Here there are some percentages for zero inflection (the remainder is inflected for *s*-genitive): *des Yangtse* 100%, *des Orinoco* 97%, *des Mississippi* 94%, *des Himalaya* 85%, *des Kongo* 73%, *des Irak* 68%, *des Iran* 65%, *des Tiber* 62%, *des Europa* 46%, *des Balkan* 23%, *des Nil* 17%, *des Neckar* 5%, *des Rhein* 0%.

Without going into more detail (there are more parameters interacting here), these percentages reveal that an increase in familiarity and an increase in similarity to native structures reduces the need for name body preservation. Thus, the name body can afford to be enlarged and altered by the -*s* suffix. Its recognizability is secured (for more detail see Zimmer 2018).

The -*s* suffix has evolved out of the formerly longer -*es* ending. To this day among many common nouns there is variation between syllabic -*es* and unsyllabic -*s*. This variation is phonologically determined (*des Tages* 'of the day' – *des Wochentags* 'of the week day', *des Landes* 'of the country' – *des Bundeslands* 'of the federal state' (for more detail see Szczepaniak 2010). Proper names are almost completely invariant. In case there is any variation, then there is variation in the form of the (older) -*s* and (in increasing numbers) zero inflection (*des vereinigten Deutschlands* > *des vereinigten Deutschland-Ø* 'of the unified Germany').[10] This introduces the process of onymic deflection (Nübling 2012). This alternation is connected with the change from -*s* > -*Ø* and not to allomorphy. Zero inflection ensures that the name body stays maximally constant. The addition of the -*s* suffix only entails small changes in phonological structure, in that (unlike -*es*) it does not add another syllable to the name. What is more, the addition of -*s* does not cancel terminal devoicing (in contrast -*es* would cancel terminal devoicing, compare *Ta[k]* vs. *Ta[g]es*).

---

10 Here, I only mention mono-inflection in passing. For more detail, consider Fritzinger (2014, 2018).

Plural inflection is similar: While the plural of the common noun *Land* 'country' is formed by umlaut, *Länder* 'countries', the proper name *Deutschland* (referring to two German states, GDR and FRG) displays variation. Plural inflection with *-s* is rare (*die beiden Deutschlands* 'the two German states') compared to zero inflection for plural (*die beiden Deutschland-Ø* 'the two German states'). The plural allomorphs of common nouns (*Burg-en* 'castles', *Berg-e* 'mountains', *Städt-e* 'cities') are not applicable to proper names. Names usually carry *-s* (or zero inflection): *die beiden Freiburg-s* 'the two Freiburgs', *Heidelberg-s* 'Heidelbergs', *Neustadt-s* 'Neustadts'. As appellative proper names contain an element that is a common noun, they should inflect like other common nouns.

However, the question is: How are appellative proper names inflected for genitive case?[11] Do they really behave exactly like common nouns as was suggested? I postulate the following hypotheses: Inflection is more likely to match that of the declension class of the formally equivalent common noun, the more a particular name in question has the status of an appellative proper name. Inflection becomes more similar to that of genuine proper names, the more the name in question has the status of a genuine proper name (with *-s* or even zero inflection). The following sections are concerned with the contrast of common nouns and appellative proper names.

Fritzinger (2014, 2018) has dealt with this in detail and thus the following sections are based on her corpus-based study of genitive inflection (DeReKo). Fritzinger's study is limited to three groups of common nouns and their formally equivalent appellative proper names, those that end in *-berg* 'mountain', *-wald* 'forest' and *-krieg* 'war' (the last one is a praxonym, i.e. a name of events). As attestations of common nouns are far more frequent than formally equivalent names, Fritzinger restricted her search to three nouns each. In contrast, Fritzinger had to include a higher number of appellative proper name types in her analysis to arrive at enough representative tokens (compare Table 7, which contains enough hits, put in brackets, as well as for common nouns and the most important appellative proper names).

---

11 In this paper, I do not analyse plural inflection, as it is more difficult than analysing genitive inflection. As names are mono-referential, they are hardly attested in the plural.

Table 7: Common nouns and appellative proper names in genitive case (Fritzinger 2014).

	-berg 'mountain'	*-wald* 'forest'	*-krieg* 'war'
(1) Common nouns	*Schuldenberg* 'mountain of debt', *Weinberg* 'vineyard', *Müllberg* 'mountain of garbage' (1248)	*Mischwald* 'mixed forest', *Nadelwald* 'conifer forest', *Regenwald* 'rain forest' (1801)	*Atomkrieg* 'nuclear war', *Stellungskrieg* 'static warfare', *Luftkrieg* 'aerial warfare' (1010)
(2) Appellative proper names	*Donnersberg, Fichtelberg, Feldberg, Falkenberg,* etc. (708)	*Schwarzwald, Odenwald, Frankenwald, Thüringer W.* etc. (4694)	*Golfkrieg, Sechstagekrieg, Vietnamkrieg* (4793)

Fritzinger (2014) contrasted the group ending in *-berg* with proper names (e.g. *Heidelberg*). Table 8 displays the percentages of the respective genitive singular endings.

Table 8: Genitive endings of nouns and names ending in *-berg* (Fritzinger 2014).

	Example	*-es*	*-s*	*-Ø*
(1) Common nouns	*des [Müll-]berg-_* 'the garbage mountain's'	41%	59%	–
(2) Appellative proper names	*des [Feld]berg-_* 'Feldberg mountain's'	30%	67%	3%
(3) Genuine proper names	*des [...] Heidelberg-_* 'Heidelberg's'	–	28%	72%

The result for the genuine proper name *Heidelberg* (3) is very clear: As expected, only the short *-s* ending (28%) and the Ø ending are attested (72%).[12] In contrast, only 30% of appellative proper names of the type *[Feld]berg* (2) take the long *-es* ending, whereas 67% are inflected with the short *-s* ending and only 3% have

---

[12] This comparison is problematic. When *Heidelberg* occurs in front, it always features *-s* (*Heidelbergs Innenstadt* 'Heidelberg's inner city'). This construction is ungrammatical when an appellative proper name is accompanied by a determiner (**des Schwarzwalds Berge* 'the Black Forest's mountains'). When searching for attestations of *Heidelberg* that are inflected in the genitive, where the preceding determiner (+attribute) is secondary, inflection of only one component of the NP dominates for other reasons (of mono-inflection): As the determiner is clearly showing genitive case, *Heidelberg* is mostly not inflected for genitive case, consider *des mittelalterlichen Heidelberg-Ø* 'the Heidelberg of the Middle Ages' (see Fritzinger 2014: 38–42).

zero inflection. Common nouns (1) heavily fluctuate and feature the long ending as well as the short ending, while completely refraining from zero inflection.

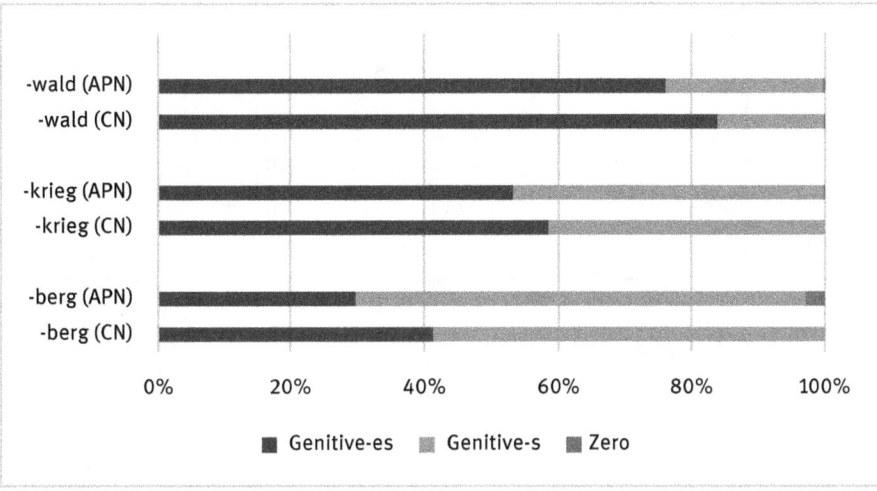

Figure 1: Genitive inflection of appellative proper names (APN) in comparison to common nouns (CN), adapted from Fritzinger (2018: 401).
For the absolute numbers see Table 7.

Figure 1 contains the results for compounds of common nouns (CN) and for appellative proper names (APN) with -*wald* 'forest', -*berg* 'mounain', and -*krieg* 'war' (genuine proper names are excluded). Clearly, appellative proper names take part in *es/s* allomorphy almost as strongly as common nouns, while they still have a tendency (10%) to adhere to the more onymic short ending more strongly. Zero inflection hardly enters the picture, mostly among types involving -*berg* (3%), where in comparison, there are mostly attestations of short -*s* endings (67%). As phonological criteria, in particular concerning the word-final sound, influence the distribution of *es/s*, differentiating these is important for the respective nouns. In all cases, the ratio of variation concerning inflection attaching to appellative proper names is slightly more onymic.

In having a brief look at particular names, individual outliers can be observed. It is not to be expected that all appellative proper names ending in -*wald* or -*berg* share the same degree of familiarity. Table 9 contains examples showing that occasionally, there are only very few tokens.

Table 9: Singular appellative proper names and their genitives (adapted from Fritzinger 2014).

	Examples	-es		-s		-ø	
		abs.	%	abs.	%	abs.	%
-wald '-forest'	des Böhmerwald-_	110	89	14	11	–	–
	des Frankenwald-_	194	83	40	17	–	–
	des Schwarzwald-_	712	66	360	34	2	< 1
	des Odenwald-_	741	63	440	37	6	0,5
	des Steigerwald-_	135	57	100	43	–	–
-berg '-mountain'	des Fichtelberg-_	21	52	19	48	–	–
	des Donnersberg-_	40	37	61	57	6	5
	des Schneeberg-_	48	28	121	71	2	1
	des Hirschberg-_	11	23	37	77	–	–
	des Feldberg-_	20	17	95	82	1	1

All in all, the way appellative proper names are inflected shows a tendency towards the genuine proper name paradigm. Yet, the strength of this tendency differs from item to item. Thus, the grammatical properties of individual names can either be closer to the paradigm of common nouns or to that of genuine proper names. Generally, appellative proper names behave more similarly to their formally equivalent common nouns. It has to be assumed that appellative proper names are affected by a gradual, chronological and successive dissociation process.

3 Conclusion

In German, the preservation of the name body, which constitutes a special grammar for names, is highly functional (see Nübling 2017). Genuine proper names no longer unfold any semantic meaning. They are reduced to the materiality of the name body and thus as a pure label they serve a direct reference. Hence, the bodies of proper names are maximally preserved.

Appellative proper names are hybrids, they are less proprialised and their semantics is only partially active. Thus, they can be distinguished from genuine proper names by the lower degree of name body preservation. Grammatically, appellative proper names share stronger correlations with common nouns. Five criteria are presented, one semantic criterion and four grammatical criteria, that are combined in Figure 2.

	1) common nouns	2) app. proper names	3) genuine proper names	
1. Semantic match	yes *Industrieland* 'industrial country'	yes *Münsterland* 'Münsterland region'	possibly *Deutschland* 'Germany'	none *Westerland* (city)
2. Loss of headedness/referential gender	Right-hand Head Rule *die Großstadt* (f.) 'the city (f.)'	Right-hand Head Rule *die Wartburg* (f.) 'the Wartburg castle (f.)'	Referential gender (loss of headedness) *Darmstadt* (n.), *Luxemburg* (n.)	
3. Behaviour of determiners	Regular determiners (full paradigm) *der/ein Weinberg* 'the/a vineyard'	definite determiner only (without paradigm) *der Feldberg* 'Feldberg mountain'	"expletive det."/ classifier possible *die Sahara* 'the Sahara desert'	no "determiner"/classifier *Deutschland* 'Germany'
4. Ellipsis	frequent	limited	impossible	
5. Inflection	full inflection	full inflection	uniform and reduced inflection or even deflection	

Figure 2: Criteria to distinguish three types of noun classes between common nouns and proper names.

As all three noun classes constitute a diachronic continuum (as indicated by the arrow), presumably there is no clear-cut distinction between the categories but rather a scalar transition (consider the dashed separating line).

On closer inspection it can be observed that it is genuine proper names that need to be divided into subtypes. Harweg (1983) falsely subsumes the first subtype under the appellative proper name class, because of the semantic match of the meaning of one of its parts and its referent (*Deutschland* 'Germany, country of the Germans', *England* 'England, country of the English', *Darmstadt* 'Darmstadt (city)'). Yet, concerning every other grammatical criterion, this subtype acts just like a genuine proper name: *Darmstadt* has long lost its head, in that neuter gender is assigned referentially (due the fixedness of neuter gender for cities) and it is no longer assigned by the Right-hand Head Rule (which would assign feminine gender). *Darmstadt* is no longer accompanied by a determiner and it cannot be used in elliptic structures as none of its parts can be omitted (**wir fahren von Bier- nach Darmstadt* 'we drive from Bier- to Darmstadt'). If there were two cities with the name *Darmstadt*, then they would be referred to as *die beiden Darmstadts* 'the two Darmstadts'. It is no longer possible to refer to them as **Darmstädte*, which indicates plural by umlaut (as is still possible e.g. for *zwei Hegelstraßen* 'two Hegel streets' or *Bismarkplätze* 'Bismarck Squares').

The second subtype involving semantic mismatches is undisputed (*Westerland* as a name for a city on Sylt). When an appellative proper name evolves into a genuine proper name, the appellative part gradually loses its referential potential. Hence the loss of semantic meaning and reference of one of the parts cannot be considered a prerequisite of genuine proper name status. This would imply that either the object or the noun referring to this object would have to change, which is not necessarily the case. However, if these changes did take place, they would not result in a renaming, as the referential connection to the formally equivalent noun has been lost (consider *Schwarzwald* 'Black Forest' as a name for a mountain range). Consequently, the transition of appellative proper name to genuine proper name is not only manifested by a semantic mismatch of the last element of the respective name. Grammar has already reacted to the semantic bleaching and is much more precise in assigning name status.

The fifth criterion, (reduced) inflection, which is directly connected to the preservation of the name body, should be unchallenged. The loss of a determiner (3) serves to stabilise the name body and it adds contour to the name body. Furthermore, the loss of a determiner is another characteristic distinguishing nouns, which are obligatorily accompanied by a determiner, from names. The same applies to the solidification of names, which becomes evident in their decreasing ability to be part of elliptic structures (4): Morphological structures and hierarchies are levelled and fused to form a holistic whole. The name becomes increasingly solid and is treated as a simplex. Ultimately, the grammatical gender itself is no longer adhering to the right-most element of the name but it is assigned (from outside) in accordance to the respective name class (2).

Gradually, the evolving name loses its appellative structures and morphosyntactic characteristics. It freezes and solidifies to form a pure sequence of sounds. Former morphological boundaries expire, the name body becomes more amorphous, so that formerly complex (appellative) proper names live on as simplex and monolithic units (e.g. *Bottrop* < *borg-thorpe* 'borough-thorpe, castle-village'). In how far graphematic means, such as a decrease in hyphenation or the use of spaces, correlate with name status needs to be determined by way of more detailed research (consider Fritzinger 2014, 2018 for some initial indications). Compare typical appellative proper names such as *Johannes Gutenberg-Universität, Emmi-Pikler-Haus, Pfälzer Wald, Gut-Hirten-Kirche* as opposed to genuine proper names such as *Neuenhaus, Oberhausen, Greifswald, Neunkirchen* without hyphens or spaces. This results in an increase in visual cohesion and solidification.

References

Ackermann, Tanja. 2018. *Grammatik der Eigennamen im Wandel. Diachrone Morphosyntax der Personennamen im Deutschen.* Berlin & Boston: De Gruyter.
Bergmann, Pia. 2018. *Morphologisch komplexe Wörter: Prosodische Struktur und phonetische Realisierung* (Studies in Laboratory Phonology 5). Berlin: Language Science Press.
DeReKo = Institute for the German Language. 2017. *Mannheim German Reference Corpus / Corpora of Written Contemporary Language 2017-II Archive* (Released on 1.10.2017). Mannheim: Institute for the German Language. www.ids-mannheim.de/DeReKo (checked 25/10/2017).
DECOW = German web corpus (checked 25/10/2017). https://corporafromtheweb.org/decow16.
Fahlbusch, Fabian & Damaris Nübling. 2014. Der Schauinsland – die Mobiliar – das Turm. Das referentielle Genus bei Eigennamen und seine Genese. *Beiträge zur Namenforschung* 49(3). 245–288.
Fritzinger, Julia. 2014. Des Feldberges, des Feldbergs oder des Feldberg-Ø? Zur onymischen (De-)Flexion genuiner Gattungseigennamen im Spannungsfeld zwischen Appellativa und Eigennamen. Mainz: Johannes Gutenberg-Universität Mainz MA Thesis.
Fritzinger, Julia. 2018. Während des Golfkrieges, des Golfkriegs oder des Golfkrieg? Gattungseigennamen im Spannungsfeld zwischen Eigennamen und Appellativa. In Rita Heuser & Mirjam Schmuck (eds.), *Sonstige Namenarten – Stiefkinder der Onomastik*, 383–406. Berlin & Boston: De Gruyter.
Harweg, Roland. 1983. Genuine Gattungseigennamen. In Manfred Faust & Peter Hartmann (eds.), *Allgemeine Sprachwissenschaft, Sprachtypologie und Textlinguistik*, 157–171. Tübingen: Gunter Narr Verlag.
Harweg, Roland. 1997. *Namen und Wörter. Aufsätze* – vol. I. Bochum: Brockmeyer.
Harweg, Roland. 1998. *Namen und Wörter. Aufsätze* – vol. II. Bochum: Brockmeyer.
Kempf, Luise. 2008. Zu meiner Errett- und Erhaltung. Synchrone und diachrone Studie zur Ellipse von Wortteilen im Deutschen. Mainz: Johannes Gutenberg-Universität Mainz MA Thesis.
Kempf, Luise. 2010. In Erober: vnd Plünderung der Statt: Wie die Ellipse von Wortteilen entstand. *Beiträge zur Geschichte der deutschen Sprache und Literatur* 132(3). 343–365.
Nowak, Jessica & Damaris Nübling. 2017. Schwierige Lexeme und ihre Flexive im Konflikt: Hör- und sichtbare Wortschonungsstrategien. In Nanna Fuhrhop, Renata Szczepaniak & Karsten Schmidt (eds.), *Sichtbare und hörbare Morphologie* (Linguistische Arbeiten 565), 113–144. Berlin & Boston: De Gruyter.

Nübling, Damaris. 2005. Zwischen Syntagmatik und Paradigmatik: Grammatische Eigennamenmarker und ihre Typologie. *Zeitschrift für Germanistische Linguistik* 33(1). 25–56.

Nübling, Damaris. 2012. Auf dem Wege zu Nicht-Flektierbaren: Die Deflexion der deutschen Eigennamen diachron und synchron. In Björn Rothstein (ed.), *Nicht-flektierende Wortarten*, 224–246. Berlin & Boston: De Gruyter.

Nübling, Damaris. 2017. The growing distance between proper names and common nouns in German: On the way to onymic schema constancy. In Tanja Ackermann & Barbara Schlücker (eds.), *The morphosyntax of proper names*. [Special issue]. *Folia Linguistica* 51(2). 341–367.

Nübling, Damaris. 2018. Vom Oden- in den Schwarzwald, von Eng- nach Irland? Zur Abgrenzung von Gattungseigennamen und reinen Eigennamen. In Rolf Bergmann & Stefanie Stricker (eds.), *Namen und Wörter. Theoretische Grenzen – Übergänge im Sprachwandel*, 11–32. Heidelberg: Winter.

Nübling, Damaris. 2020. Die Bismarck – der Arena – das Adler. The emergence of a classifier system for proper names in German. In Johanna Flick & Renata Szczepaniak (eds.), *Walking on the grammaticalization path of the definite article. Functional main and side roads* (Studies in Language Variation 23), 227–249. Amsterdam & Philadelphia: John Benjamins.

Nübling, Damaris, Rita Heuser & Fabian Fahlbusch. 2015. *Namen. Eine Einführung in die Onomastik*. 2nd edn. Tübingen: Narr.

Paul, Hermann. 1917/1968. *Deutsche Grammatik, Teil III: Flexionslehre* – vol. II. Tübingen: Max Niemeyer.

Schäfer, Roland. 2015. Processing and querying large web corpora with the COW14 architecture. *Proceedings of Challenges in the Management of Large Corpora (CMLC-3)*. 28–34.

Schäfer, Roland & Felix Bildhauer. 2012. Building large corpora from the web using a new efficient tool chain. *Proceedings of the Eight International Conference on Language Resources and Evaluation (LREC'12)*. 486–493.

Schmuck, Mirjam. 2017. Movierung weiblicher Familiennamen im Frühneuhochdeutschen und ihre heutigen Reflexe. In Johannes Helmbrecht, Damaris Nübling & Barbara Schlücker (eds.), *Namengrammatik* (Linguistische Berichte – Sonderheft 23), 33–58. Hamburg: Buske.

Steche, Theodor. 1927. *Die neuhochdeutsche Wortbiegung unter besonderer Berücksichtigung der Sprachentwicklung im 19. Jahrhundert*. Breslau: Hirt.

Szczepaniak, Renata. 2010. Während des Flug(e)s/des Ausflug(e)s? German Short and Long Genitive Endings between Norm and Variation. In Alexandra Lenz & Albrecht Plewnia (eds.), *Grammar between norm and variation*, 103–126. Frankfurt: Peter Lang.

Zimmer, Christian. 2018. *Die Markierung des Genitiv(s) im Deutschen. Empirie und theoretische Implikationen von morphologischer Variation* (Reihe Germanistische Linguistik 315). Berlin & Boston: De Gruyter.

Iker Salaberri
D-marking on Basque personal names from a synchronic and diachronic perspective

Abstract: This paper offers an overview of the marking for definiteness and specificity (D-marking) on personal names and common nouns in Basque from a synchronic and diachronic perspective. In view of the fact that common nouns are increasingly and personal names decreasingly D-marked over time, it is argued that proper names as a whole are on their way towards becoming a distinct word class. Moreover, it is shown that despite this tendency specific subtypes of personal names are still D-marked in a few varieties of Modern Basque. These facts are related to the observation that proper names have a special onymic grammar.

Keywords: Basque, definiteness, grammaticalization, personal name, place name, Special Onymic Grammar.

1 Introduction

Onomastic research has traditionally focused on the etymologies of names and their relevance for language history and reconstruction. In comparison, crosslinguistic studies dealing with the synchronic and diachronic grammatical features of proper names are few (Anderson 2007; Van Langendonck 2007; Schlücker and Ackermann 2017; Handschuh 2017; Salaberri Izko 2020; see also the contributions by Caro Reina, Handschuh and Helmbrecht in this volume). Despite the growing interest in the grammatical features of proper names, detailed analyses of the morphosyntactic behaviour of PNs in less-studied non-Indo-European languages are still scarce. The goal of this paper is to provide an overview of the marking of definiteness and specificity (D-marking) on proper names (PNs) and common nouns (CNs) in Basque, a language isolate spoken in the northeast of Spain and the southwest of France.

More specifically, this study aims to answer the following questions: (1) are there differences concerning D-marking on Basque personal names and CNs?;

Iker Salaberri: University of the Basque Country, Paseo de la Universidad 5, 01006 Vitoria-Gasteiz, Spain, iker.salaberri@ehu.eus

(2) does D-marking on personal names vary depending on personal name subclass?; and (3) does D-marking on personal names change historically and diatopically? These questions raise an issue, though: The comparatively brief history of Basque written records provides an incomplete view of the changes undergone by D-marking on PNs and CNs, since these were already underway by the time the first large written texts came to light in the 16th century. An attempt has been made to compensate this incomplete view by reconstructing previous stages of the definite article on the basis of proposals by Azkue (1923), *Orotariko euskal hiztegia* (1: 2–3), Manterola (2006, 2015), De Rijk (2008) and Etxeberria (2014).

The paper is organized as follows: Section 2 presents an account of D-marking on CNs and PNs in Basque from a synchronic perspective, followed in Section 3 by a discussion concerning D-marking on CNs and PNs in Basque from a diachronic perspective. Finally, Section 4 presents the conclusions and analyzes potential implications for the study of names.

2 D-marking in Basque from a synchronic perspective

2.1 Common nouns

The main D-marker in Modern Basque is the definite article -*a* (plural -*ak*, from -*a* and the plural suffix -*k*).[1] This element is a bound morpheme that is attached to the end of noun phrases in compliance with group inflection, which means that -*a(k)* can be found next to various elements including nouns, adjectives and nominalized verbs (Trask 2003: 119–120). In Standard Basque, -*a(k)* fulfills the basic functions typically associated with definite articles: anaphoric use for referring to something previously mentioned in discourse and non-anaphoric use for referring to something that the speaker assumes to be known to the hearer, for instance by means of general knowledge (Dryer 2013). These uses are illustrated in (1a) and (1b), respectively:

---

1 The D-marker form -*e* (plural -*(e)k*), instead of -*a(k)*, is common in many dialects (Zuazo 2008: 239–241), cf. example (9b).

(1) a. *Roblesene-ko-ek,   Luzaide-ko   etxe-a-z*
Roblesene-RM-ERG.PL   Luzaide-RM   house-DM-INS
bestalde,   ba-z-u-te-n   Auritz-en   etxe-a,
besides   AFF-3PST-have-PL-PST   Auritz-LOC   house-DM
herri-an   sar-tze-an   eskuin, lehenbizi-ko etxe-a
town-LOC   enter-NMLZ-LOC   right   first-RM   house-DM
'Those from the Roblesene house owned, besides a house in Luzaide, a(nother) house in Auritz, (which is) the first house on the right (of the road) when entering the town.' (Aintziburu & Etxarren 2002: 40)
b. *Eguzki-a-k   argi-tzen   du*
sun-DM-ERG   shine-IPFV   AUX
'The sun is shining.' (De Rijk 2008: 266)

As pointed out by Trask (2003: 118–119), however, *-a(k)* is used more broadly than just for conveying definiteness. This morpheme can, in fact, indicate indefinite specific (2a) and generic (2b) referents, mark complement clauses with a generic interpretation (2c) and even attach to vocatives (2d) (see Manterola 2015: 74–76 for a comprehensive overview):[2]

(2) a. *Emazte-a   d-a-uka-t*
wife-DM   3PRS-PRS-have-1SG
'I have a wife.' (Trask 2003: 119)
b. *Ardo   gorri-a   nahiago   dut*
wine   rosé-DM   prefer   AUX
'I prefer rosé wine.' (Trask 2003: 119)
c. *Merezi   du   entsaio   ttiki   bat   egi-te-a*
be.worth   AUX   test   little   INDF   do-NMLZ-DM
'It is worth doing a little test.' (Zabala 2000: 98)
d. *Ume-a-k,   etorr-i   hona!*
child-DM-PL   come-PFV   hence
'Come here, children!' (Trask 2003: 121)

Due to the wide range of functions of *-a(k)*, it seems inaccurate to label this morpheme as a definite article. The broad terms "D-element" (Himmelmann 1997: 6–7), "(pattern of) D-marking" (Handschuh 2017: 499) and "D-marker"

---

[2] In general terms, bare nouns and bare adjectives are disallowed in Standard Basque when they constitute arguments of the verb (Etxeberria 2014: 335).

(Salaberri Izko 2020: 560), which encompass markers of specificity in addition to definiteness, seem more appropriate. Thus, henceforth the morpheme -*a(k)* will be referred to as "D-marker".

The variety of uses of the Standard Basque D-marker can be described in diachronic terms with regard to the grammaticalization cline from demonstrative to noun marker, for which different grammaticalization paths have been proposed. Following Greenberg (1978: 61), -*a(k)* might be described as transitioning from Stage I (definite article) to Stage II (specific article) because it encompasses uses associated with the former (definite) as well as the latter (specific, generic and non-referential). The Standard Basque D-marker cannot, however, be regarded as a Stage III (nominal marker) element. This would imply that -*a(k)* "no longer has any synchronic connection with definiteness or specificity" (Greenberg 1978: 69), which is not the case, as argued above.

On the basis of Lyons' (1999: 337) implicational hierarchy of the functional expansion of D-elements, it can be claimed that Standard Basque resembles Italian and Portuguese (Lyons's Stage 3) in that -*a(k)* can occur in definite (1a–b), generic (2b) and possessive constructions (3a–b) while it differs from Catalan and Greek (Lyons's Stage 4) in that D-marking on proper names is disallowed (see Section 2.2):

(3) a. *Neure   aspaldi-ko   amets-a-k   ahaztu-rik*
       1SG.INT.GEN very.old-RM   dream-DM-PL   forget-PTCP
       bizi   naiz   mundu-tik   urruti
       live   AUX   world-ABL   far.away
       'Having forgotten my very old dreams, I live far away from the world.'
       (De Rijk 2008: 698)
    b. *Neke-z   egin   nuen   nire   bide-a*
       difficulty-INS   make.PFV   AUX   1SG.GEN   way-DM
       'I made my way with difficulty.' (De Rijk 2008: 757)

A further grammaticalization pathway of D-elements is provided by Lehmann (2015: 59), which fuses the formal and functional dimensions of change. According to this proposal, Basque can be claimed to have a so-called "affixal article" (Lehmann's Stage 4). Moreover, at this stage both definite and specific use of the D-element is possible (Lehmann 2015: 41), as has been argued to be true of Basque -*a(k)*.

To summarize, Standard Basque can be described —regardless of the model chosen for description— as a language where the grammaticalization of the D-marker -*a(k)* has reached an advanced stage such that it enables not only definite, specific and generic uses, but also additional uses. This element has not (yet), however, become a noun marker.[3]

The claims made so far apply for Standard Basque and for most Basque dialects. By contrast, the easternmost varieties of Basque (Zuberoan, eastern Low Navarrese and the now extinct Salazarese and Roncalese) are slightly different in the sense that nouns and adjectives lack D-marking under specific conditions:[4] First, direct object nouns (*behi* 'cow', *sagar* 'apple') are left bare in order to favor a non-definite (specific or generic) reading (4a–b). Second, non-D-marked predicative adjectives (*polit* 'nice', *eder* 'pretty') indicate constant properties (4c–d):

(4) a. *Bortü-an   ikus-i   dit   behi*
       mountain-LOC   see-PFV   AUX   cow
       'I saw cows in the mountain.' (Etxeberria 2014: 344)
   b. *Zer   agi-tü   da?   Sagar   ebats-i   dü*
       what   happen-PFV   AUX   apple   steal-PFV   AUX
       'What happened? S/he stole apples.' (Etxeberria 2014: 344)
   c. *Jestu   hori   ez   d-a   polit*
       gesture   that   NEG   3PRS-be   nice
       'That gesture is not nice.' (Manterola 2011: 76)
   d. *Zühaintze   hori-en   oro-ren   lili-a-k*
       tree   those-GEN   all-GEN   flower-DM-PL
       eder   d-ir-a
       pretty   3PRS-be.PL-PRS
       'The flowers of all those trees are pretty.' (Manterola 2011: 76)

Crucially, attaching *-a(k)* to the nouns and adjectives in question would trigger a definite reading in these dialects. It should be noted that bare nouns and adjectives like those in (4a–d) would be ungrammatical in Standard Basque and most other dialects. Therefore, *-a(k)* seems to behave more like a definite article in eastern varieties whereas in Standard Basque and the remaining dialects it is between a definite article and a specific article. On the basis of historical attestations (see Section 3 below) and in line with the grammaticalization cline of D-

---

3 The grammaticalization of the plural D-marker *-ak* involves a number of additional nuances that are of little concern to the present discussion (see Michelena 1970, Martínez Areta 2009 and Manterola 2015 for details).
4 The classification of Basque dialects in this paper follows Zuazo (2008). For an extensive overview of the uses of *-a(k)* in eastern Basque varieties, see Camino (2009) and literature cited therein.

elements outlined above, it has been argued that Zuberoan, eastern varieties of Low Navarrese, Salazarese and Roncalese represent an earlier stage of the language (Michelena 1964a; Manterola 2011; Etxeberria 2014).

It should be mentioned that peripheral (i.e., western and eastern) dialects of Basque treat a number of CNs as inherently definite, thus disallowing D-marking: *etxe* 'house', *ugazaba* 'boss, employer, owner' —this one only in western dialects—, *errege* 'king' and *erregina* 'queen'. Some kinship terms also fall into this group: *ama* 'mother', *aita* 'father', *amona ~ amama ~ amatxi ~ amañi* 'grandmother', *aitona ~ aitaita ~ aitatxi ~ aitañi* 'grandfather', *izeba ~ izeko* 'aunt' and *osaba* 'uncle' (Azkue 1923: 278; Manterola 2015: 78).

2.2 Proper names

So far it has been argued that D-marking of CNs is a very widespread phenomenon in Standard Basque as well as in most Basque dialects, albeit to a lesser degree, in the easternmost varieties (Zuberoan, eastern Low Navarrese, Salazarese, Roncalese). By contrast, proper names provide a different picture: As a general rule, *-a(k)* is not attached to proper names. This is particularly true of personal names (5a–b) and most place names (5c–d):

(5) a. *Jacob   izan   nuen   maite   eta   ez   Esau*
       Jacob   have.PFV   AUX   love   and   NEG   Esau
       'I loved Jacob and not Esau.' (De Rijk 2008: 1064)
    b. **Jacob-a   izan   nuen   maite   eta   ez   Esau-a*
       Jacob-dm   have.PFV   AUX   love   and   NEG   Esau-DM
       'I loved Jacob and not Esau.'
    c. *D-a-ki-da-n-ez,*   *Madril-en*   *bizi*   *zen*
       3PRS-PRS-KNOW-1SG-REL-INS   Madrid-LOC   live   AUX
       'As far as I know, s/he lived in Madrid.' (De Rijk 2008: 1132)
    d. **D-a-ki-da-n-ez,*   *Madril-a-n*   *bizi*   *zen*
       3PRS-PRS-know-1SG-REL-INS   Madrid-DM-LOC   live   AUX
       'As far as I know, s/he lived in Madrid.'

Apparent counter-examples include instances of deproprialization (6a), place names which historically resulted from CNs and bear an older, no longer productive, organic *-a* (6b), place names that occur with classifiers (6c) and personal names ending in *-a* due to phonological reasons (6d). For example, *Jordiak* in (6a) cannot be considered a personal name, but rather a common noun (Schlücker and Ackermann 2017: 311). Place names such as *Bizkaia* in (6b), which originally derived from CNs, reflect a preonymic state which does not display the full synchronic properties of a PN. Moreover, (6c) illustrates instances of place names occurring with classifiers, i.e., with lexemes which also function as CNs and which indicate the class of the topographical entity referred to by the place name. Place names accompanied by classifiers are marked and, therefore, display less properties typical of proper names (Van Langendonck 2007: 210; Nübling et al. 2015: 105–106). This suggests that in (6c) the presence of D-marking, which in Modern Basque most frequently follows CNs, is due to the classifier (*uharteak* 'the islands') rather than the proper name (*Azoreak* 'the Azores').

(6) a. *Jordi-a-k   kale-an   gura   ditugu*
       Jordi-DM-PL   street-LOC   want   AUX
       'We want the Jordis (i.e., the men called Jordi) out on the street (i.e., to be free).' (Ahal Dugu Arrasate Podemos Mondragon 2017)
   b. *Bizkai-a-k   mendi-z   apaindu-ri-ko   gaztelu   zar*
      Bizkai-DM-ERG   mountain-INS   decorate-PTCP-RM   castle   old
      bat   d-irudi
      INDF   3PRS-look.like
      'Bizkaia (orig. 'the mountain range') looks like an old castle decorated with mountains.' (Enbeita 1974: 114)
   c. *Azore-a-k   eta   Kanaria-r   uharte-eta-ko*
      Azores-DM-PL   and   Canary-GNT   island-LOC.PL-RM
      kostalde-a-k   animalia hil-ez   bete   ditu
      coast-DM-PL   animal   dead-INS   fill   AUX
      'S/he filled the Azores and the coasts of the Canary Islands with dead animals.'
      (Sarasola et al. 2009)
   d. *Trenpü   gaitz-ean   d-ü-gü   Pierra*
      disposition   ill-LOC   3PRS-have-1PL Pierre
      'Pierre is ill-disposed.' (Coyos 2013: 86)

The same is generally true of non-prototypical place names such as names of mountains (7a), seas (7b) and oceans (7c):

(7) a. *Agintari   musulman-a-k   Orreaga-ko   bide-tik*
       ruler   Muslim-DM-ERG   Orreaga-RM   pass-ADL
       zeharka-tu   zituen   Pirinio-a-k
       cross-PFV   AUX   Pyrenees-DM-PL
       'The Muslim ruler crossed the Pyrenees through the Orreaga pass.'

(Olaizola 2006: 83)
b. *Mediterraneo-a   zeharka-tu,   eta   basamortu-ko*
   Mediterranean-DM   cross-PFV   and   desert-RM
   haize-a-k   Bartzelona   har-tu   du
   wind-DM-ERG   Barcelona   reach-PFV   AUX
   'The desert wind has crossed the Mediterranean and reached Barcelona.'
   (Ordoñez)
c. *Antarktika   orain-go   India-r   Ozeano-a-ren*
   Antarctica   now-RM   India-GNT   Ocean-DM-GEN
   leku-an   egon-go   zen
   area-LOC   be-FUT   AUX
   'Antarctica would have been located in the present-day Indian Ocean area.' (Sarrionandia 2001: 281)

Proper names can, in turn, take other modifiers including adjectives (8a-b), possessive adjectives (8c)[5] and titles (8d). Here, *-a(k)* attaches to the modifiers since, as mentioned above, in Standard Basque the D-element cannot attach to proper names (Trask 2003: 161):

(8) a. *Gal-du   egin   nuen   Bilbo   maite-a*
       lose-PFV   do.PFV   AUX   Bilbo   beloved-DM
       'I lost (my) beloved Bilbo.' (Tolkien 2004: 331)
   b. *San Mames   berri-a-ri   alokairu   txikiegi-a*
      San Mames   new-DM-DAT   rent   too.small-DM
      jar-tze-a   aurpegira-tu   diote   aldundi-a-ri
      grant-NMLZ-DM   accuse-PFV   AUX   provincial.council-DM-DAT
      'The provincial council has been accused of granting the new San Mames (name of a football stadium) too low a rent (to be payed)' ("San Mames Berriari")
   c. *Jon   gure-a*
      John   1PL.GEN-DM
      'Our John' (Trask 2003: 162)
   d. *Karlos   bostgarren-a*
      Charles   fifth-DM
      'Charles V' (Trask 2003: 129)

---

[5] Compare *Jon gurea* 'our John' (8c), which is rare and possible only in formal speech, with *gure Jon* 'our John', which is common in casual speech and does not allow for the presence of a D-marker (Trask 2003: 162).

Finally, word-final -*a* in some first names in Zuberoan Basque (see (6d) above) can be argued not to be a D-marker, but rather an epenthetic vowel inserted in order to avoid certain word-final consonants, as pointed out by Michelena (1959: 8–10). In summary, none of the examples mentioned so far can be claimed to represent genuine exceptions to the rule that PNs are not D-marked.

Nevertheless, D-marked PNs can be found in a number of non-standard varieties of Modern Basque. For example, Salaberri Zaratiegi (2007: 393) reports D-marking of settlement names (9a) in Bortziriak Basque, a sub-dialect of High Navarrese. The same observation has been made for Lekunberri (9b) (Central Basque) and Goizueta (sub-dialect of High Navarrese) (9c):

(9) a. *Doneztua (< Doneztebe-a) / Ittuna (< Ituren-a or Iturin-a) / Oiza (< Oiz-a) / Urroza (< Urroz-a)*
'Doneztebe / Ituren / Oiz / Urroz (names of hamlets)' (Salaberri Zaratiegi 2007: 393)
b. *Lekunberri-e*
Lekunberri-DM
'Lekunberri (name of a city)' (Jimeno Jurio 1991: 74)
c. *Pedro-a-ren-a*
Peter-DM-GEN-DM
'Pedroarena (lit. the house of the Peter) (name of a farmhouse)'
(Salaberri and Zubiri 2009: 820)

These examples of D-marked settlement names (9a–b) and D-marked personal name within an oikonym (9c) do not fall into the same group as *Bizkaia* in (6b) above for a number of reasons. First, place names which derived historically from CNs and bear an older, no longer productive, organic -*a* such as *Azpeitia*, *Bizkaia* and *Iruñea* are systematically inflected like CNs. This implies that the erstwhile D-element must disappear in front of locative case endings (De Rijk 2008: 58), cf. *Iruñea* in Table 1:

Table 1: Inflection of common nouns, appellative proper names and genuine proper names in Basque (adapted from De Rijk 2008: 58).[6]

Case	Common noun	Appellative proper name	Genuine proper name
Absolutive	*hiri* 'city'	*Iruñea*	*Donostia*
Locative	*hiri-an*	*Iruñe-an*	*Donostia-n*
Ablative	*hiri-tik*	*Iruñe-tik (*Iruñea-tik)*	*Donostia-tik*
Adlative	*hiri-ra*	*Iruñe-ra (*Iruñea-ra)*	*Donostia-ra*
Terminative	*hiri-raino*	*Iruñe-raino (*Iruñea-raino)*	*Donostia-raino*
Directive	*hiri-rantz*	*Iruñe-rantz (*Iruñea-rantz)*	*Donostia-rantz*

By contrast, D-marking of *Doneztebe, Ituren, Lekunberri*, etc. seems to be a more sporadic and optional phenomenon confined to the corresponding local dialect. The second reason why *Azpeitia, Bizkaia* and *Iruñea* should be distinguished from *Doneztebe, Ituren* and *Lekunberri* is their etymology: The former quite uncontroversially originate from the the common nouns *a(i)z-be(h)eiti* 'foot of the mountain', *bizkai* 'mountain range' and *iruñe* 'city' (Salaberri Zaratiegi 1993: 190–192; Salaberri and Zaldua 2020: 92–94) whereas this is not necessarily true of the latter. *Doneztebe*, for example, is the result of Basque rendering of *Done Eztebe* 'Saint Stephen', clearly a Romance hagionym (Salaberri Zaratiegi 2004: 160).

In addition to place names such as *Doneztebe, Ituren* and *Lekunberri*, a few hypocoristics are accompanied by *-a(k)* in Leitza Basque, another central variety. Interestingly, in this variety D-marking conveys a familiar meaning and is employed only with female hypocoristics (10a–b). Male hypocoristics, on the other hand, cannot take *-a(k)* (10c). Note that *-txo* in (10a–c) constitutes the diminutive ending.

(10) a. *Dominika   eta   Joakina-txo-a   etorr-i   die*
       Dominique   and   Joakina-DIM-DM   come-PFV   AUX
       kontu-k   esa-te-a
       matter-PL   tell-NMLZ-ADL
       'Dominique and little Joakina came to tell about some matters.'
       (Salaberri Izko 2020: 573)
   b. *Maria-txo-a   oso   gaizki   men   d-a*
       Mary-DIM-DM   very   bad   EVDL   3PRS-PRS

---

6 See Nübling (this volume) and references cited therein for the terms *appellative proper names* and *genuine proper names*.

'They say little Mary is doing very badly.' (Salaberri Izko 2020: 573)
c. *Austin-txo (*Austin-txo-a) / Fermin-txo (*Fermin-txo-a) / Ramon-txo (*Ramon-txo-a)*
'Little Austin / Little Fermin / Little Ramon' (Salaberri Zaratiegi 2009: 215)

These D-marked proper names (9a–c, 10a–b) are only a few exceptions to the general rule that proper names are not D-marked, which also applies to the dialects. Moreover, these exceptions come from varieties spoken in northern and north-western Navarre. These include Bortziriak, Goizueta, Leitza and Lekunberri Basque. These varieties do not allow for systematic D-marking of proper names. Rather, the degree to which this is possible seems to depend on the dialect and proper name class (or subclass).[7]

2.3 Summary

To sum up, in Standard Basque, *-a(k)* has a wide range of uses, including but not restricted to the definite, specific, generic and non-referential ones. This diversity of meanings has been related to the advanced grammaticalization of the D-markers. By contrast, D-marking of personal and place names is disallowed despite a number of apparent counterexamples.

In contrast to Standard Basque, the easternmost dialects Zuberoan, eastern Low Navarrese, Roncalese and Salazarese allow bare nouns and bare adjectives as arguments of the verb. Such bare arguments allow for a specific or generic reading. In turn, attaching *-a(k)* to nouns and adjectives favors a definite interpretation. Accordingly, *-a(k)* has been argued to behave rather like a definite article in the easternmost varieties. In accordance with the grammaticalization path of demonstrative to noun marker, these dialects can be claimed to represent a more conservative stage of the language.

In addition, Basque dialects differ from Standard Basque with respect to the marginal tolerance of D-marking on some personal and place names. More specifically, a few varieties from northern and north-western Navarre allow for D-marking of specific subtypes of place names and personal names, subject to

---

[7] Trask (2003: 162) reports D-marking of the deity name *Jesus* in *Jesusaren bihotza* 'Jesus's heart', with no reference to dialect. However, this should rather be considered a remnant of the use of *-a* with deity names (see Section 3.2 for details).

dialectal variation. These are, however, rare exceptions that are far from being systematic.

Cross-linguistically, the presence of D-marking with CNs and its absence with proper names is not common (Handschuh 2017: 499; Salaberri Izko 2020: 561). Nevertheless, this pattern is widespread in standard varieties of European languages such as French, German, Italian, Spanish and Swedish, as opposed to some of their dialects, which allow for D-marking of personal names. In this respect, some scholars talk about a Standard Average European feature (Seiler 2019; Caro Reina this volume). Basque has a striking parallel with standard European languages, which raises the question of whether this could also be due to areal influences on a continental level. This issue will be further discussed in Section 3.

3 D-marking in Basque from a diachronic perspective

3.1 Common nouns

The D-element -*a* developed from a distal demonstrative pronoun **(h)a(r)* in preliterary Basque, during the Middle Ages (Azkue 1923: 269; *Orotariko euskal hiztegia* 1: 2–3; Manterola 2015: 3).[8] This is a very common path of grammaticalization of D-elements (Greenberg 1978; Kuteva et al. 2019: 137–139). Some of the earliest attestations of D-marked common nouns are found in the list of Basque words attributed to the 12th-century Poitevin monk Aymeric Picaud, who probably made the list during his pilgrimage to Santiago de Compostela (11a). By contrast, other common nouns from the same list are bare (11b). Unfortunately, due to lack of context it is impossible to determine the precise reading of -*a* in these examples. It is true, however, that most D-marked CNs in the list are count nouns whereas those without -*a* are mass nouns:

---

8 In texts up to the 19th century, -*a(k)* competes for grammaticalization with the D-elements -*au*, -*ori*, -*on* and -*ok* among others, some of which are believed to derive from the proximal and medial demonstratives *hau* and *hori* (*Orotariko euskal hiztegia* 3: 295–300; 13: 565–570; Manterola 2015). This competition does not seem to be directly related to the present discussion.

(11) a. *Eche-a / iaon-a / andre-a / belaterr-a / eregui-a*
'the house / the lord / the lady / the clergyman / the king'
(Michelena 1964b: 50)
b. *Orgui / ardum / aragui / araign / gari*
'Bread / wine / meat / fish / wheat' (Michelena 1964b: 50)

Already in late-medieval (15th-century) texts, the D-marker -*a* seems to behave as a definite article, at least in the sense of referring to something that the speaker assumes to be known to the hearer from context or general knowledge. Thus, in (12a) -*a* accompanies the noun *lur* 'earth'. In (12b), which is part of a couplet dedicated to the high-ranking officer of a city militia, -*a* attaches to the noun phrase *condestable jaun* 'lord high-ranking officer'. In (12c), which belongs to a letter where among other topics money issues are discussed, the CN *colectore* 'tax collector' is D-marked:

(12) a. *Lur[r]-a-c   d-a-c[a]r   og[i]*
       earth-DM-ERG   3PRS-PRS-BRING   bread
       'The earth brings bread.' (ca. 1425; Michelena 1964b: 58)
   b. *Condestable   jaun-á   ar   bizate   anáie*
       high.ranking.officer lord-DM   take   AUX   brother
       'May he (the king) take the high-ranking officer as a brother.' (1494; Reguero 2019: 61)
   c. *Bay-t-a-tor   sey   florin   et   t[er]cio*
       REL-3PRS-PRS-come   six   florin   and   third
       bat   yl-ean   rebati-çe-ra   colectore-a-ri
       one   month-LOC   collect-NMLZ-ABL   tax.collector-DM-DAT
       'Who comes once a month to collect six florins and one third from the tax collector.' (1416; Monteano 2015: 149)

In addition, shortly after the beginning of the literary period in the mid-16th century common nouns that are arguments of the verb can be left bare in order to favor an indefinite specific (13a) and indefinite non-specific (generic) (13b) reading. In (13a) the noun phrases *spiritu ithobat* 'a troubled spirit' and *begui* 'eye' are both direct objects of the verb *eman vkan draue* 's/he gave them to them'. The first direct object is accompanied by the numeral *bat* 'one', which behaves like an indefinite article.[9] The second direct object is the bare noun

---

9 The grammaticalization of the numeral *bat* as an indefinite article is related to the present discussion (see Manterola 2012a or details). However, this topic is beyond the scope of this study.

begui 'eye', which has an indefinite and specific interpretation (Michelena 1970: 76–77).[10] Similarly, in (13b) the direct object is the bare noun *gorroto* 'hatred', which has a non-specific generic interpretation. These examples suggest that, unlike in the modern standard language, indefinite specific and indefinite non-specific common nouns that constitute arguments of the verb did not need to be D-marked in 15th- and 16th-century Basque.

(13) a. *Eman   vkan   draue   Iainco-a-c   spiritu   itho=bat:*
       give   PFV   AUX   God-DM-ERG   spirit   troubled=INDF
       eta   begui,   ikus   ez=tezate-n-çát
       and   eye   see   NEG=AUX-SUB-SUB
       'God gave them a troubled spirit; and eyes, so that they would not see.' (Leizarraga 1990 [1571]: 844)
   b. *Mondr[a]goe-ri   har-tu   deusat   gorroto*
       Mondragón-DAT   take-PFV   AUX   hatred
       'I have come to hate Mondragón (lit. I have taken hatred towards Mondragón).' (ca. 1586–1598; Michelena 1964b: 76)

The same is true of non-referential common nouns (14a–b). The common nouns *bildurr* 'fear' and *quibel* 'back' are the direct objects of the verbs *egosçi* 'to pour' and *eguin* 'to do, make' in (14a) and (14b), respectively. These nouns are quite likely non-referential because they are part of idiomatic expressions (Manterola 2012b: 192–193), and yet they are bare. This contrasts with Modern Standard Basque, where non-referential common nouns that are arguments of the verb must be D-marked (see Section 2.1).

(14) a. *Bildurr   egosç-i*
       fear   pour-PFV
       'To frighten (lit. to pour fear (into someone))' (Landuchio 1958 [1562]: 67)
   b. *Quibel   egui=oc   ecach-a-ri*
       back   make=AUX   storm-DM-DAT
       'Turn your back on a storm (lit. make back on a storm).' (1596; Lakarra 1996: 358)

Another property of 15th-16th-century Basque is that predicative adjectives can be left bare in order to indicate constant properties (15a–b). In (15a) *on* 'good' is

---

10 It should be noted that example (13a) is part of a translation. Therefore, translation effects on D-marking cannot be ruled out.

a predicative adjective, just like *hon* 'good' and *leyal* 'faithful' in (15b). None of these adjectives bears the D-element *-a(k)*. Again, this is unlike the modern language, where all nominal predicates that are arguments of the verb require D-marking, apart from a few exceptions that are of no concern to the present discussion (see De Rijk 2008: 19–20 for details).

(15) a. Arma   escudu-a-c   on   d-ir-a   dardo-s
     weapon shield-DM-PL good 3PRS-be-PRS arrow-INS
     'Weapons and shields are good (when accompanied) by arrows.' (ca. 1558–1598; Arriolabengoa 2006: 225)

   b. Ar-çen   dut   neure   semazte-ren Johana
     take-IPFV AUX 1SG.INT.GEN wife-GEN Johana
     hau   eta   hon   eta   leyal   n-ay-ça-ca-n,
     this   and   good   and   faithful 1SG-PRS-be-3SG-SUB
     ala   d-u-da-n-a-s   eta
     whether 3PRS-have-1SG-REL-DM-INS and
     d-u-que-da-n-a-s
     3PRS-have-POT-1SG-REL-DM-INS
     'I take this Johana as my wife, so that I may be good and faithful to her (and provide for her) with whatever I have and might have.' (1506; Reguero 2019: 69)

There is to date no extensive quantitative analysis of the frequency of D-marking on common nouns in 17th-20th-century Basque. Moreover, as shown by Manterola (2015: 383–457), many variables influence the likelihood for common nouns to be D-marked in 16th-century Basque, which include the speech act performed by the utterance, sentence polarity, type of clause, transitivity of the verb, and case. Despite these variables, a look at the distribution of *-a(k)* in 17th-20th-century texts indicates that the frequency of D-marking on common nouns with low referentiality increases over time (16a–b). In (16a) the D-element *-a* attaches to the noun phrase *ardo gorri* 'rosé wine', even though the referent is indefinite and non-specific (generic). Moreover, in (16b) the predicative adjective *on-a* 'good', which indicates a constant property, is likewise D-marked. Therefore, these examples suggest that, by the 18th-19th centuries, D-marking of CNs has extended to environments which were left bare in 15th-16th-century Basque.

(16) a. Cer   ardo   nai   dute, jaun-a-c?   Ardo   gorri-a
     which wine want AUX lord-DM-PL wine rosé-DM

'Which wine do the gentlemen want? Rosé wine.' (Anonymous 2004 [1868])

b. *Ecen*   *Jainco-a-c*   *crea-tu*   *du-en*   *guci-a*   *on-a*
   since   God-DM-ERG   create-PFV   aux-REL   all-DM   good-DM
   d-a
   3PRS-PRS
   'Since everything that God has created is good.' (Haraneder 1990 [1740]: 433)

To summarize this section, the behaviour of the D-element *-a(k)* in 15th-16th-century Basque has been retained in the easternmost varieties of the modern language: D-marking indicates definiteness (12a–c) and common nouns that constitute arguments of the verb can be left bare in order to favor an indefinite specific (13a), indefinite non-specific (generic) (13b) or non-referential (14a–b) reading. Furthermore, non-D-marked adjectives indicate constant properties (15a–b). This leads to the following conclusions: First, *-a(k)* can best be described as a definite article in 15th-16th-century Basque, as in the modern easternmost varieties. Second, in accordance with the grammaticalization path from demonstrative to noun marker outlined in Section 2.1, *-a(k)* has developed indefinite specific, non-specific (generic) and non-referential readings over time (see 16a–b). Third, the modern easternmost varieties of Basque have largely preserved the range of uses of *-a(k)* from 15th-16th-century Basque. Therefore, these varieties can be claimed to represent an earlier stage of the language. There is consensus among scholars concerning these three issues (among others: Azkue 1923: 265; Lafon 1954: 114; Michelena 1970: 292–293; Martínez Areta 2009; Manterola 2015).

3.2 Proper names

The D-marker *-a(k)* is attested with deity names (17a–b) at least as early as the 16th century.[11] In (17a) the D-element *-a(k)* attaches to the deity name *Jesus*

---

11 Even earlier cases of D-marked PNs in Basque can involve medieval attestations of place names bearing endings such as *-ha*, *-zaha*, *-a*, *-aga*, *-eta* and *-en*. However, the D-element status of these endings is not always straightforward. Moreover, many of these place names are in fact frozen forms consisting of one or more D-marked common nouns, which is why they reflect a preonymic state similar to (6b) above. Accordingly, proper names from before the 16th century have been left out of the discussion (see Manterola 2015: 165–239 and Salaberri Zaratiegi 2015: 42–43 for details).

whereas in (17b) it accompanies the hagionym *Iandone Peri* 'Saint Peter'. Most cases of D-marked proper names in 16th-20th-century Basque involve names of holy beings and are attested in western texts.

(17) a. *Ce   egun-e-ta-n   circuncida-u   zàn,   eta   imin-i*
       which   day-LV-INN-LOC   circumcize-PFV   AUX   and   take-PFV
       eutsen   Jesus-a-en   icen-a?
       AUX   Jesus-DM-GEN   name-DM
       'On which day did He circumcize and take the name of Jesus?'
       (Olaetxea 1787 [1763]: 21)
    b. *Ni   becatari-au   confess-etan   nachaco [...]*
       1SG   sinner-this   confess-IPFV   AUX
       Iandone   Peri-a-ri
       Saint   Peter-DM-DAT
       'I, who am a sinner, confess to Saint Peter.' (1596; Michelena 1955: 57)

This is further illustrated by (18a–b), which concern the hagionym *Sant Franciscu* 'Saint Francis' and the deity name *Satanas* 'Satan' respectively:

(18) a. *Aita   Sant   Franciscu-a-ren   ordea-co-a*
       Father   Saint   Francis-DM-GEN   order-RM-DM
       'From the order of Father Saint Francis' (first half of the 16th century; Omaetxebarria 1948: 310)
    b. *Olgueeta   madarica-tu   orr-e-quin*
       game   damn-PTCP   that-LV-COM
       d-a-uca-la   satanas-a-c   itsu-tu-ta
       3PRS-PRS-have-SUB   Satan-DM-ERG   blind-PFV-PTCP
       Euscal-errij-eta-co   gente-ric   gueijeen-a
       Basque-country-LOC.PL-RM   people-PART   most-DM
       'That Satan has blinded most people in the Basque Country with that damned game.' (Santa Teresa 1986 [1816]: 164)

Despite this dialectal distribution of D-marked names in 16th-20th-century Basque, similar examples involving the deity name *Jesus* are less frequently found in farther eastern texts (19a–b). Example (19a) is written in the Gartzain variety, which falls into High Navarrese, whereas (19b) belongs to Altsasu Basque, a transitional variety between Western Basque-Central Basque-High Navarrese. In geographic terms, both examples stem from northern-northwestern Navarre, which resembles Modern Basque with respect to the distribution of D-marked proper names (see Section 2.2). Nevertheless, the fact that D-marked deity names are attested in western texts (17–18) as well suggests that D-marking of PNs could geographically have been more widespread than it is in the modern language.

(19) a. *Jesus-a   jaio   eta   am=ori   bakaŕik,*
       Jesus-DM   be.born   and   mother=that   alone
       su-a-ren   egi-te-ko   e=tza-uka-n
       fire-DM-GEN   make-NMLZ-RM   NEG=3PST-have-PST
       eguŕ-ik
       wood-part
       'When Jesus was born His mother was alone, she did not have wood to light a fire.' (Azkue 1990 [1925]: 1078)
   b. *Josepe   ta   Maria   beren   Jesus-a-kin   yoan*
       Joseph   and   Mary   3PL.GEN   Jesus-DM-COM   go.PFV
       ziran   Belen-era
       AUX   Bethlehem-ADL
       'Joseph and Mary went to Bethlehem together with their Jesus.' (Azkue 1990 [1925]: 1109)

Nonetheless, personal names bearing *-a(k)* are the exception rather than the rule in 16th-20th-century Basque. Indeed, first names (20a), hagionyms (20b) and deity names (20c) usually do not take the D-marker:

(20) a. *Ethorr-i   da   Ioannes   çue-tara   iustitia-z-co*
          come-PFV   AUX   John   2PL-ADL   justice-INS-RM
          bide-a-z,   eta   ez=tuçue   hura   sinhets-i
          path-DM-INS   and   NEG=AUX   3SG   believe-PFV
          'John came to you along the path of justice, and you did not believe him.' (Leizarraga 1990 [1571]: 41)
   b. *Horr-en   gaign-ian   erran=du*
      that-GEN   on-LOC   say=AUX
      S. Augustin-ec   hitz   hon=bat
      Saint Augustine-ERG   word   good=INDF
      'Saint Augustine has spoken a good word about that.' (Tartas 1672: 30)
   c. *Dabiltza   billá-tzen   mundu-tic   quen-tze-co*
      PROG   look.for-IPFV   world-ABL   wipe.off-NMLZ-RM
      Jesus
      Jesus

'They are looking to wipe Jesus off the earth.' (Lizarraga Elkanokoa 1983 [1821]: 79)

In view of the dialectal variation concerning the D-marking of proper names, the frequency with which -a(k) attaches to names might be determined by drawing on quantitative corpus analysis. Table 2 shows the results of an analysis of onymic D-marking in 16th-20th-century Basque. The data have been collected from the Corpus of Basque Classics (*Euskal klasikoen corpusa*), an 11,9-million-word database which comprises texts written between 1545 and 1985.

Table 2: Rate of D-marked names of saints and deities (1545–1985).

Onym	D-marking	No D-marking
Jesus, Iesus, Yesus, Yosus	42/29,529 (0.14%)	29,487/29,529 (99.86%)
(San, Iandone) Pedro, Peru, Petri, Peri	5/5,974 (0.08%)	5,969/5,974 (99.92%)
(San, Iandone) Franzisko, Franzisku, Pranzisko	10/2,500 (0.4%)	2,490/2,500 (99.6%)
(San, Iandone) Ignazio, Inazio, Iñazio	3/1,986 (0.15%)	1,983/1,986 (99.85%)
(San, Iandone) Antonio, Anttonio	1/1,532 (0.07%)	1,531/1,532 (99.93%)
Satan, Satanas	39/1,026 (3.8%)	987/1,026 (96.2%)
Total:	100/42,547 (0.24%)	42,447/42,547 (99.76%)

First, it should be noted that the names of saints and deities shown in Table 2 are the only ones in the corpus that bear -a(k). Second, Table 2 confirms the observations made so far: -a(k) attaches only to a reduced number of names of saints and deities, and D-marking of these proper names is marginal since its frequency is lower than 5%.

This raises the question of whether D-marking of names of saints and deities is a diachronically stable phenomenon. Figure 1 captures the frequency of D-marked vs. non-D-marked proper names in 50-year intervals. Despite being a marginal phenomenon, the frequency of D-marked proper names has decreased over time: After a near-3% peak during the 1650–1699 period, the rate of names bearing -a(k) gradually diminishes. This suggests that deity names bearing the D-element as in *Jesusaren bihotza* 'Jesus's heart' (Trask 2003: 162) and which exceptionally occur in Modern Standard Basque cannot be linked to D-marking of proper names in modern dialects (see examples 9–10, Section 2.2). That is, D-marking of proper names is exceptional both in modern dialects and in Modern

Standard Basque. However, in the former it is still a productive rule (subject to the conditions mentioned in Section 2.2), whereas in the latter it is not.

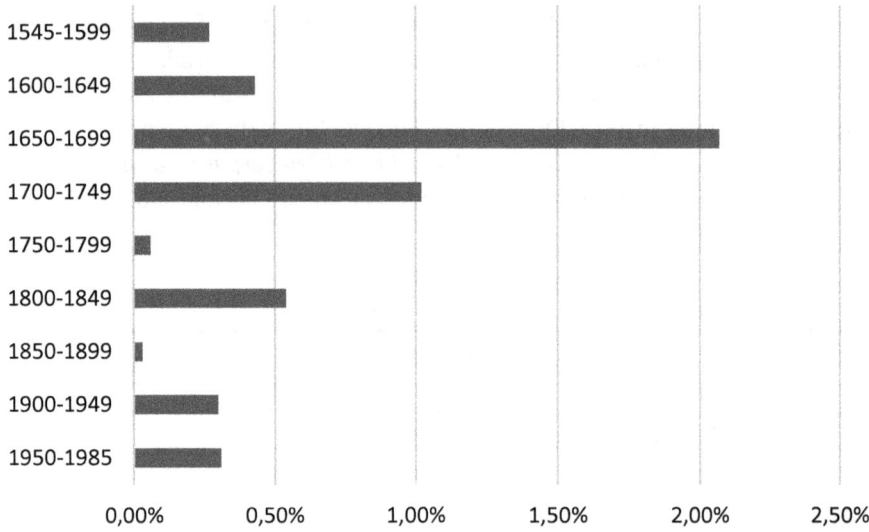

Figure 1: Rate of D-marked names of saints and deities (1545–1985).

Table 3 captures the rate of onymic D-marking across 16th-20th-century Basque dialects. The findings reveal that D-marked proper names are not particularly frequent in any of the Basque dialects. However, we can observe a slightly higher tendency for peripheral (western and eastern) dialects to attach -*a(k)* to names of saints and deities. In any case, the general frequencies as well as the differences among dialects are too low to draw any conclusions in this respect.

Table 3: Rate of D-marked names of saints and deities according to dialect (1545–1985).

Dialect	D-marked proper names	Non-D-marked proper names
Western Basque	43/5,706 (0.75%)	5,663/5,706 (99.25%)
Central Basque	23/21,675 (0.11%)	21,652/21,675 (99.89%)
High Navarrese	5/1,686 (0.3%)	1,681/1,686 (99.7%)
Navarrese-Lapurdian	26/12,850 (0.2%)	12,824/12,850 (99.8%)
Eastern Basque	3/630 (0.48%)	627/630 (99.52%)
Total:	100/42,547 (0.24%)	42,447/42,547 (99.76%)

To summarize, a few hagionyms and deity names bearing -a(k) are attested in 16th-20th-century Basque. However, these represent only a handful of cases out of all names attested in Basque texts. Moreover, the corpus analysis shows that, even for the few names to which -a(k) can attach, this is a marginal phenomenon that never exceeds 5% of all cases. The differences between dialects are too small to draw any valid conclusions, but a slight diachronic tendency can be observed, according to which D-marked proper names become less frequent over time. In Modern Standard Basque, D-marking of proper names is ungrammatical save for some exceptions.

These generalizations do not entirely match the observations made in Section 2.2 concerning onymic D-marking in modern dialects. More specifically, hypocoristics bearing -a(k), which occur in a few dialects, do not seem to be attested in 16th-20th-century Basque. However, this can be due to reasons unrelated to the topic at hand. For instance, most of written Basque literature up to the 19th century is religious in nature, which may have conditioned the absence of hypocoristics in the corpus. In fact, a closer look at secular texts reveals that hypocoristics are indeed occasionally D-marked in 16th-20th-Basque (21a–d):[12]

(21) a. *Vechi-a*[13]   *de*   *Olayz*
 Little.Peter-DM   of   Olaitz
 'Little Peter of Olaitz' (1514; Salaberri Zaratiegi 2018: 715)
 b. *Martin-ico*   *fijo*   *de Johan-to-a*
 Martin-DIM   son   of John-DIM-DM
 'Little Martin, son of Little John' (1518; Salaberri Zaratiegi 2009: 124)
 c. *Petri-cho-a*
 Peter-DIM-DM
 'Little Peter' (1560; Salaberri Zaratiegi 2009: 222)
 d. *Miguel-to-a / Petri-tu-a / Betri-tto-a / Petri-tto-a / Petri-cho-a / Petri-to-a*
 'Little Michael / Little Peter / Little Peter / Little Peter / Little Peter / Little Peter' (1753–1847; Satrustegi 1961: 213)

---

12 For hypocoristics bearing -*a* in Medieval Basque, see Salaberri Zaratiegi (2009: 166, 207–211).
13 *Vechi* is a hypocoristic form of *Bet(r)i, Pet(i)ri* 'Peter' (Salaberri Zaratiegi 2018: 715). The sequence of sound changes that derived *Vechi* from *Bet(r)i, Pet(i)ri* presumably involves loss of the rhotic followed by palatalization of the voiceless stop (/t/ > /tʃ/ or /c/, which is represented in the spelling as <ch>) and voicing of the word-initial plosive (/p/ > /b/, which is represented in the spelling as <v>).

These D-marked hypocoristics (21a–d) are interesting for two reasons: First, they represent a diachronic link between 16th-20th-century Basque and modern dialects (10a–b). Second, most of them are male hypocoristics whereas in the modern varieties that allow for onymic D-marking, -*a(k)* only attaches to female hypocoristics (see Section 2.2). Nevertheless, female hypocoristics bearing the D-element are also attested in 16th-20th-century Basque (22a–b):

(22) a. *Gorainci   ene   espos   Maria-to-a-ri*
        greetings   1SG.GEN   wife   Mary-DIM-DM-DAT
        secula-n   yçan-en   duda-n-a-ri
        always-LOC   have-FUT AUX-REL-DM-DAT
        'Greetings to my wife, Little Mary, whom I shall always have.' (1662; Michelena 1964b: 144)

   b. *Ni   bederen,   ene   bizi-a-ren   ema-te-ra*
      1SG   at.any.rate   1SG.GEN   life-DM-GEN   give-NMLZ-ADL
      n-a-go   ene   Margarita-ño-a-ren   alde
      1SG-PRS-be 1SG.GEN   Margaret-DIM-DM-GEN   for
      'I, at any rate, am (willing) to give my life for my Little Margaret.' (Duvoisin 1987 [1832]: 46)

In short, both male and female hypocoristics can bear -*a(k)* in 16th-20th-century Basque, although in absence of corpus attestations it is difficult to determine how frequent this phenomenon is. By contrast, in modern dialects only female hypocoristics (10a–b) can be D-marked (Salaberri Izko 2020: 573). Again, this suggests that D-marking of proper names has become more restricted over time.

A final example of D-marked proper names involves city names, at least two instances of which (23a–b) are attested in 16th-20th-century literature:

(23) a. *Donapalayo-a,   la   ville   de   S. Palais*
        Donapaleu-DM   the   city   of   Saint Palais
        en   basse   nauarre
        in   low   Navarre
        'Donapaleu, the town of Saint-Palais in Low Navarre' (Pouvreau 17th century: 58)

   b. *Lore   ta   berar   Erbeste-co   oec,   beren*
      flower   and   herb   foreign-RM   DEM.PL   3PL.GEN
      jaioterri-co   icen-a-quin   dei-tzen   dituzte
      native.country-RM   name-DM-COM   call-IPFV   AUX
      Oñati-a-n

Oñati-DM-LOC
'Those foreign flowers and herbs are called by their native names in Oñati (name of a town).' (Iztueta 1847: 160)

Similar to hypocoristics, the attestations of city names bearing -*a(k)* (23a–b) represent a link between 16th-20th-century Basque and modern dialects (see Section 2.2). They also indicate that D-marking on city names must have been geographically more widespread in previous centuries: Example (23a) is from Navarrese-Lapurdian whereas example (23b) is from Central Basque. By contrast, as mentioned above, most examples of D-marked city names in Modern Basque stem from the High Navarrese variety.

These facts raise a number of questions: It is not clear why only deity names, hypocoristics and city names are D-marked in 16th-20th-century Basque whereas other proper name classes are left bare. Regarding deity names, one can argue, in line with Caro Reina (2020), that they are more identifiable, agentive and context-independent than other proper name classes. To the extent that -*a(k)* marks referents that are inferable from context or general knowledge in 16th-20th-century Basque (see Section 3.1), one could argue that the D-element attaches to names that are known to the hearer through particular religious knowledge or devotion, as suggested by Azkue (1990 [1925]: 842).

As far as hypocoristics are concerned, they are a subclass of personal names, which have been argued to be more animate than other proper name classes (Nübling et al. 2015: 102). In this sense, animacy is believed to correlate with additional properties including agentivity and familiarity, for instance with regard to their grammatical behaviour (Schlücker and Ackermann 2017: 317). Therefore, D-marking of hypocoristics in Basque can serve to indicate that they are more familiar or readily identifiable to the hearer. This proposal fails to explain, however, why -*a(k)* does not attach to other personal name subclasses.

In the same vein, city names have been argued to involve a high degree of human involvement in comparison to other place names such as names of deserts (Van Langendonck 2007: 140). Thus, animacy might explain why city names are accompanied by -*a(k)* while other place name classes are left bare in dialects of 16th-20th-century Basque as well as in Modern High Navarrese.

A second question involves the development of D-marking on common nouns vs. proper names in 16th-20th-century Basque: In accordance with the grammaticalization path of D-elements outlined in Section 2.1, -*a(k)* has innovated indefinite specific, non-specific (generic) and non-referential readings over time. Consequently, common nouns have become increasingly D-marked in the history of Basque. By contrast, D-marking of proper names seems to have decreased. It should be noted, however, that this observation is based on the analysis of a restricted and very infrequent phenomenon that is only partially attested in the sources. Therefore, this observation should be taken with a grain of salt. Nonetheless, it is clear that D-marking of proper names has not increased at the same rate as D-marking of common nouns.

If we consider, as the data suggest, that D-marking of proper names has decreased, then this asymmetric development is not supported by the traditional grammaticalization of D-elements according to which -*a(k)* would extend to new contexts. Instead, it acquired new functions and at the same time lost others. In their recent analysis of the grammaticalization cline from demonstrative to noun marker, Szczepaniak and Flick (2020: 4) speak against a strictly linear progression of grammaticalization. Alternatively, they advocate that each new function acquired by the former demonstrative can follow a distinct diachronic succession. In view of the Basque data, the observation can be added that some functions of the D-element can subside in the course of grammaticalization.

The third and final question concerns the possible motivation(s) behind the unexpected development of Basque -*a(k)*. In this respect, two explanations are possible: First, increased D-marking of CNs as opposed to decreased D-marking of PNs means that names and CNs have become more different over time. This can be indicative that proper names are developing into a distinct word class. The diverging grammatical behaviour of proper names in Basque has been previously observed regarding another property: case markers of spatial relations (locative, adlative, ablative) are shorter for proper names than for common nouns (Stolz et al. 2014; Haspelmath 2021: 13–14). Therefore, it might well be that the reduction in D-marking is contributing to distinguishing proper names from common nouns. In general terms, one could argue that the absence of D-marking on PNs in Modern Standard Basque is due to the special onymic grammar of names.

The second explanation involves contact. Modern Basque is similar to major European languages with definite articles such as French, German, Italian, Spanish and Swedish in the sense that bare PNs are distinctive of the standard variety whereas proper names are D-marked in a few dialects. This is not quite the case for 16th-20th-century Basque, which means that Basque has become more similar to the major European languages. Accordingly, one can argue that the absence of D-marking on proper names in Modern Standard Basque is a Standard Average European feature. In fact, both the distinct grammatical behaviour of proper names and areal influences might be accountable for the Basque data.

4 Conclusions

This paper has provided an overview of D-marking on common nouns and proper names in Basque from a synchronic and a diachronic perspective. A few new insights have been gained, and previous claims have been supported with additional data. First, Modern Standard Basque disallows the use of the bound D-element *-a(k)* with proper names, as opposed to common nouns, which are nearly always D-marked when they constitute arguments of the verb. By contrast, the easternmost dialects (Zuberoan, Eastern Low Navarrese, Salazarese and Roncalese) exhibit bare common nouns and adjectives that are verbal arguments provided that an indefinite specific or indefinite non-specific (generic) reading is possible. Moreover, a few varieties from northern and northwestern Navarre (Bortziriak, Goizueta, Leitza and Lekunberri Basque) marginally attach *-a(k)* to female hypocoristics and city names.

Second, the behaviour of *-a(k)* with common nouns in 15th-16th-century Basque is very similar to modern eastern varieties. Therefore, in accordance with the grammaticalization path of demonstrative to noun marker, eastern dialects of Modern Basque can be claimed to represent an earlier stage of the language. In turn, *-a(k)* occasionally attaches to deity names and, to a lesser extent, hypocoristics and city names in some 16th-to-20th-century dialects. These attestations represent a diachronic link to D-marked proper names in modern varieties and further indicate that D-marking of proper names has become more restricted both functionally and diatopically. In general terms, we can observe that D-marking of common nouns has increased and spread to new contexts whereas the use of the D-element *-a(k)* with proper names has probably decreased and is absent from some contexts in which it was used before (male hypocoristics, hagionyms). The sequence of changes seems to have unfolded in the following manner:

Stage 1 (before grammaticalization of the D-element *-a(k)*):
  CN-Ø, PN-Ø
Stage 2 (after grammaticalization of *-a(k)* and up to the 20th century):
  CN-*a(k)*, PN-*a(k)* (only in some very restricted contexts)
Stage 3 (Modern Standard Basque, most modern dialects):
  CN-*a(k)*, PN-Ø

This uneven development of D-marking on CNs and PNs does not quite correspond to the traditional view of grammaticalization of D-elements, according to which the former demonstrative spreads to new contexts over time. Rather, the data suggest that each new use acquired by the D-element follows a distinct diachronic succession, and that some functions —particularly onymic D-marking— are reversible.

Third, the lack of D-marking on PNs in Modern Standard Basque, considered along with the distinct onymic markers for spatial cases (locative, adlative, ablative), suggests that proper names are in the process of becoming a distinct word class in Basque. This process seems to have unfolded more or less in parallel to the grammaticalization of the D-element -*a(k)*. From a cross-linguistic perspective, this different behaviour might be a reflection of the special grammar of names (Anderson 2007: 127; Nübling et al. 2015: 64). Finally, the presence of D-marking on proper names in a few dialects vs. its absence in the standard variety implies that Basque has become more similar to major European languages concerning definite articles. Accordingly, the Basque data might be an areal Standard Average European feature.

Acknowledgments

I would like to thank Patxi Salaberri Zaratiegi, the editors and three anonymous reviewers for useful comments and constructive criticism. Thanks are also due to Bittorio Lizarraga for providing data from Leitza Basque. Financial support from the Spanish Ministry of Science, Innovation and Universities (through grant PGC2018-098995-B-I00, research group *Diacronía de la animacidad: aproximación tipológica hacia el origen de marcas animadas*, head researcher Prof. Dr. Iván Igartua) is likewise gratefully acknowledged. Any remaining errors are my sole responsibility.

Abbreviations

1, 3	first person, third person
ABL	ablative
ADL	adlative
AFF	affirmative
AUX	auxiliary
COM	comitative
CN	common noun
DAT	dative

DIM	diminutive
DM	D-marker
ERG	ergative
EVDL	evidential
FUT	future
GEN	genitive
GNT	gentilic
INDF	indefinite
INN	indeterminate
INS	instrumental
INT	intensive
IPFV	imperfective
LOC	locative
LV	linking vowel
NEG	negation
NMLZ	nominalizer
PART	partitive
PFV	perfective
PL	plural
PN	proper name
POT	potential
PROG	progressive
PRS	present
PST	past
PTCP	participle
REL	relativizer
RM	relational marker
SG	singular
SUB	subordinator

References

Ahal Dugu Arrasate Podemos Mondragon. 2017. *Jordiak kalean gura ditugu. Errepresio politikoaren aurka, askatasun demokratikoa aldarrika dezagun gaur Arrasaten 19:30etan Herriko plazan* [We want the Jordis out on the streets. Against political repression, let us proclaim democratic freedom today in Arrasate at 19:30 at the town square]. Facebook. https://hi-in.facebook.com/PodemosArrasate/posts/jordiak-kalean-gura-ditugu-errepresio-politikoaren-aurka-askatasun-demokratikoa-/1417130501737576/ (checked 17/12/2021).

Aintziburu, Angel & Jean-Baptiste Etxarren. 2002. *Luzaiden gaindi* [Across Luzaide]. Donostia: Elkar.
Anderson, John M. 2007. *The grammar of names*. Oxford: Oxford University Press.
Anonymous. 2004 [1868]. Yracasdearen escugarria [Teacher's handbook]. In Rosa M. Pagola, Itziar Iribar & Juan J. Iribar (eds.), *Bonaparte ondareko eskuizkribuak – Fondo Bonaparte*. Arrasate/Mondragón: Deustu University.
Arriolabengoa, Julen. 2006. *Ibarguen-Cachopín kronika: Edizioa eta azterketa* [The Ibarguen-Cachopín chronicle: Edition and analysis]. Vitoria-Gasteiz: University of the Basque Country dissertation.
Azkue, Resurrección María de. 1923. *Morfología vasca*. Bilbo: La Gran Enciclopedia Vasca.
Azkue, Resurrección María de. 1990 [1925]. *Cancionero popular vasco*. 3 vol. 3rd edn. Bilbo: Euskaltzaindia.
Camino, Iñaki. 2009. Mugako hiztun eta aldaerak ipar-mendebaleko Zuberoan [Frontier speakers and varieties in north-west Zuberoa]. *Fontes Linguae Vasconum* 111. 153–218.
Caro Reina, Javier. 2020. Differential object marking with proper names in Romance languages. In Luise Kempf, Damaris Nübling & Mirjam Schmuck (eds.), *Linguistik der Eigennamen* (Linguistik, Impulse & Tendenzen 88), 225–258. Berlin: De Gruyter.
Caro Reina, Javier. this volume. The definite article with personal names in Romance languages.
Coyos, Jean-Baptiste. 2013. *Zubererazko istorio, alegia eta ipuin barre-irrigarri* [Humorous stories, allegories and tales in Zuberoan (Basque)]. Bilbo: Euskaltzaindia.
De Rijk, Rudolf P. G. 2008. *Standard Basque: A progressive grammar. Volume 1: The grammar*. Cambridge MA & London: MIT Press.
Dryer, Matthew S. 2013. Definite articles. In Matthew S. Dryer & Martin Haspelmath (eds.), *The world atlas of language structures online*. Leipzig: Max Planck Institute for Evolutionary Anthropology. http://wals.info/chapter/37 (checked 24/12/2021).
Duvoisin, Jean. 1987 [1832]. *Baigorriko zazpi liliak* [The seven flowers of Baigorri]. Donostia: Elkar.
Enbeita, Balendin. 1974. *Nere apurra* [My bit]. Tolosa: Auspoa.
Etxeberria, Urtzi. 2014. Basque nominals: From a system with bare nouns to a system without. In Ana Aguilar-Guevara, Bert Le Bruyn & Joost Zwarts (eds.), *Weak referentiality*, 335–364 (Linguistik Aktuell/Linguistics Today 219). Amsterdam & Philadelphia: John Benjamins.
Euskara Institutua (ed.). 2013. *Euskal klasikoen corpusa* [Corpus of Basque classics]. Donostia: University of the Basque Country. https://www.ehu.eus/ehg/kc/ (checked 24/12/2021).
Greenberg, Joseph H. 1978. How does a language acquire gender markers? In Joseph H. Greenberg, Charles A. Ferguson & Edith A. Moravcsik (eds.), *Universals of human language. Volume 3: Word structure*, 47–82. Stanford, CA: Stanford University Press.
Handschuh, Corinna. 2017. Nominal category marking on personal names: A typological study of case and definiteness. *Folia Linguistica* 51(2). 483–504.
Handschuh, Corinna. this volume. Personal names versus common nouns: crosslinguistic findings from morphology and syntax.
Haraneder, Joanes. 1990 [1740]. *Jesu Christoren evangelio saindua* [The holy gospel of Jesus Christ]. Bilbo: Euskaltzaindia.
Haspelmath, Martin. 2021. Explaining grammatical coding asymmetries: form-frequency correspondences and predictability. *Journal of Linguistics* 57(3). 605–633.
Helmbrecht, Johannes. this volume. Proper names with and without definite articles: preliminary results.

Himmelmann, Nikolaus P. 1997. *Deiktikon, Artikel, Nominalphrase. Zur Emergenz syntaktischer Struktur* (Linguistische Arbeiten 362). Tübingen: Max Niemeyer.
Iztueta, Juan I. 1847. *Guipuzcoaco provinciaren condaira edo historia* [The legend or history of Gipuzkoa province]. Donostia: Ignacio Ramón Baroja.
Jimeno Jurio, José M. (coord.). 1991. *Nafarroako herri izendegia / Nomenclátor euskérico de población de Navarra*. Bilbo: Government of Navarre, Euskaltzaindia.
Kuteva, Tania, Bernd Heine, Bo Hong, Haiping Long, Heiko Narrog & Seongha Rhee. 2019. *World lexicon of grammaticalization*. 2nd edn. Cambridge: Cambridge University Press.
Lafon, René. 1954. Le nombre dans la déclinaison basque. *Via Domitia* 1. 111–121.
Lakarra, Joseba A. 1996. *Refranes y sentencias (1596): Ikerketak eta edizioa* [Research and edition]. Bilbo: Euskaltzaindia.
Landuchio, Nicolás. 1958 [1562]. *Dictionarium linguæ cantabricæ* [Dictionary of the Cantabrian (Basque) language]. Donostia: Gipuzkoan Provincial Council.
Lehmann, Christian. 2015. *Thoughts on grammaticalization* (Classics in Linguistics 1). 3rd edn. Berlin: Language Science Press.
Leizarraga, Joanes. 1990 [1571]. *Iesus Christ gure iaunaren testamentu berria, Othoitza ecclesiasticoen forma, Catechismea, Kalendrera, ABC edo christinoen instructionea* [New Testament of our lord Jesus Christ, Prayer for Ecclesiastes, Catechism, Christian calendar, ABC or the Instruction of Christians]. Bilbo: Euskaltzaindia.
Lizarraga Elkanokoa, Joakin. 1983 [1821]. *Koplak* [Couplets]. Bilbo: Euskaltzaindia.
Lyons, Christopher. 1999. *Definiteness*. Cambridge: Cambridge University Press.
Manterola Agirre, Julen. 2006. -*a* euskal artikulu definituaren gainean zenbait ohar [Some remarks on the Basque definite article -*a*]. *International Journal of Basque Linguistics and Philology* 40. 651–676.
Manterola Agirre, Julen. 2011. -*a* morfemaren erabilera (eza) ekialdeko euskaretan [(Lack of) use of morpheme -*a* in eastern Basque]. In Joseba A. Lakarra, Joakin Gorrotxategi & Blanca Urgell (eds.), *2nd Conference of the Luis Michelena Chair*, 71–96. Vitoria-Gasteiz: University of the Basque Country.
Manterola Agirre, Julen. 2012a. The Basque articles -*a* and *bat* and recent contact theories. In Claudine Chamoreau & Isabelle Léglise (eds.), *Dynamics of contact-induced language change* (Language Contact and Bilingualism 2), 231–263. Berlin: De Gruyter.
Manterola Agirre, Julen. 2012b. Synchronic ubiquity of the Basque article -*a*: A look from diachrony. In Urtzi Etxeberria, Ricardo Etxepare & Myriam Uribe-Etxebarria (eds.), *Noun phrases and nominalization in Basque: syntax and semantics* (Linguistik Aktuell/Linguistics Today 187), 179–206. Amsterdam & Philadelphia: John Benjamins.
Manterola Agirre, Julen. 2015. *Euskararen morfologia historikorako: Artikuluak eta erakusleak* [Towards a history of Basque morphology: Articles and demonstratives]. Vitoria-Gasteiz: University of the Basque Country PhD thesis.
Martínez Areta, Mikel. 2009. The category of number in Basque: II. Prehistorical and typological aspects. *Fontes Linguae Vasconum* 111. 249–280.
Michelena, Luis. 1955. La Doctrina Cristiana de Betolaza (1596). *International Journal of Basque Linguistics and Philology* 2. 41–60.
Michelena, Luis. 1959. Sobre -*a* en nombres vascos de persona. *Euskera* 4. 5–10.
Michelena, Luis. 1964a. *Sobre el pasado de la lengua vasca*. Donostia/San Sebastián: Auñamendi.
Michelena, Luis. 1964b. *Textos arcaicos vascos*. Madrid: Ediciones Minotauro.

Michelena, Luis. 1970. Nombre y verbo en la etimología vasca. *Fontes Linguae Vasconum* 4. 67–94.
Monteano, Peio. 2015. La carta bilingüe de Matxin de Zalba (1416). El iceberg lingüístico navarro. *Fontes Linguae Vasconum* 119. 147–173.
Nübling, Damaris. this volume. Von Heidel- nach Bamberg, von Eng- nach Irland? 'From Heidel- to Bamberg, from Eng- to Ireland?' On the delimitation of appellative proper names and genuine proper names.
Nübling, Damaris, Fabian Fahlbusch & Rita Heuser. 2015. *Namen. Eine Einführung in die Onomastik*. 2nd edn. Tübingen: Narr.
Olaetxea, Bartolome. 1787 [1763]. *Cristinauben dotrinia* [Christian doctrine]. 4th edn. Tolosa: Francisco de la Lama.
Olaizola, Jesus Mari. 2006. *Bizitza eredugarriak* [Exemplary lives]. Donostia: Elkar.
Omaetxebarria, Ignacio. 1948. El vascuence de Fray Juan de Zumárraga. *Boletín de la Real Sociedad Vascongada de Amigos del País* 4(3). 293–314.
Ordoñez, Jon. 2004. Bartzelona, hondarretan [Barcelona, buried in sand]. *Berria*. https://www.berria.eus/paperekoa/6332/033/003/2004-12-30/bartzelona-hondarretan.htm (checked 17/12/2021).
Orotariko euskal hiztegia = Michelena, Luis (ed). 1987–2005. *Orotariko euskal hiztegia / Diccionario general vasco*. 15 vol. Bilbo: Desclée de Brouwer, Euskaltzaindia, Mensajero.
Pouvreau, Silvain. 17th century. *Dictionnaire basque-français*; manuscript. Held in the Bibliothèque Nationale in Paris.
Reguero, Urtzi. 2019. *Filologiatik dialektologiara Nafarroako euskarazko testu zaharretan barrena (1416–1750)* [From philology to dialectology aross old Basque texts from Navarre (1416–1750)]. Bilbo: University of the Basque Country.
Salaberri Izko, Iker. 2020. Variable D-marking on proper naming expressions: A typological study. *Folia Linguistica* 54(3). 551–580.
Salaberri Zaratiegi, Patxi. 1993. Nafarroako hiriburuaren izenaren gainean [On the name of the capital of Navarre]. *Euskera* 38(1). 167–192.
Salaberri Zaratiegi, Patxi. 2004. *Nafarroa Behereko herrien izenak: Lekukotasunak eta etimologia* [The Names of Low Navarre villages: Records and etymology]. Iruñea: Government of Navarre.
Salaberri Zaratiegi, Patxi. 2007. Nafarroako herri izenen bukaerako txistukaria eta toponimoen arautzea [The final sibilant of village names in Navarre and the regulation of toponyms]. *Euskera* 52(1). 391–394.
Salaberri Zaratiegi, Patxi. 2009. *Izen ttipiak euskaraz* [Basque hypocoristics]. Bilbo: Euskaltzaindia.
Salaberri Zaratiegi, Patxi. 2015. *Araba / Álava: Los nombres de nuestros pueblos*. Vitoria-Gasteiz: Euskaltzaindia, Alavese Provincial Council.
Salaberri Zaratiegi, Patxi. 2018. Izen ttipiak euskaraz: Addenda, confirmanda et corrigenda [Basque hypocoristics: Additions, confirmations and corrections]. *International Journal of Basque Linguistics and Philology* 52(1–2). 713–732.
Salaberri Zaratiegi, Patxi & Juan J. Zubiri Lujanbio. 2009. Euskal deituren jatorria eta etxe izengoitiak [House nicknames and the origin of Basque surnames]. *International Journal of Basque Linguistics and Philology* 43. 819–830.
Salaberri Zaratiegi, Patxi & Luis M. Zaldua. 2020. *Gipuzkoako herrien izenak: Lekukotasunak eta etimologia* [The names of Gipuzkoan villages: Records and etymology]. Bilbo: Euskaltzaindia.

San Mames Berriari alokairu txikiegia jartzea aurpegiratu diote aldundiari [The provincial council has been accused of granting the New San Mames (name of a stadium) too low a rent (to be payed)]. *Bizkaiko Hitza* 16 October 2015. https://bizkaia.hitza.eus/2015/10/16/san-mames-berriari-alokairu-txikiegia-jartzea-aurpegiratu-diote-aldundiari (checked 20/12/2021).

Santa Teresa, Bartolome. 1986 [1816]. *Euscal-errijetaco olgueeta ta dantzeen neurrizco gatz-ozpinduba* [Moderate assorted discussion of the amusements and dances in the Basque Country]. Bilbo: Euskaltzaindia.

Sarasola, Ibon, Pello Salaburu, Josu Landa & Josu Zabaleta (eds.). 2009. *Ereduzko prosa gaur* [Current exemplary prose]. Donostia: University of the Basque Country. https://www.ehu.eus/euskara-orria/euskara/ereduzkoa/ (checked 17/12/2021).

Sarrionandia, Joseba. 2001. *Lagun izoztua* [The frozen friend]. Donostia: Elkar.

Satrustegi, Jose M. 1961. Aportación al estudio de la onomástica tradicional vasca. *Euskera* 6(1). 209–230.

Seiler, Guido. 2019. Non-Standard Average European. In Andreas Nievergelt & Ludwig Rübekeil (eds.), *'athe in palice, athe in anderu sumeuuelicheru stedi'. Raum und Sprache. Festschrift für Elvira Glaser zum 65. Geburtstag*, 541–554. Heidelberg: Winter.

Schlücker, Barbara & Tanja Ackermann. 2017. The morphosyntax of proper names: An overview. In Tanja Ackermann & Barbara Schlücker (eds.), *The morphosyntax of proper names*. [Special issue]. *Folia Linguistica* 51(2). 309–339.

Stolz, Thomas, Sander Lestrade & Christel Stolz. 2014. *The crosslinguistics of zero-marking of spatial relations* (Studia Typologica. Supplements STUF - Language Typology and Universals 15). Berlin & Boston: De Gruyter.

Szczepaniak, Renata & Johanna Flick. 2020. Introduction: Walking on the grammaticalization path of the definite article – functional main and side roads. In Renata Szczepaniak & Johanna Flick (eds.), *Walking on the grammaticalization path of the definite article* (Studies in Language Variation 23), 1–16. Amsterdam & Philadelphia: John Benjamins.

Tartas, Ioan. 1672. *Arima penitentaren occupatione devotaq* [Devout activities of the penitent soul]. Orthez: Jacques Rovyer.

Tolkien, John R. R. 2004. *Eraztunen jauna: Erregearen itzulera* [The lord of the rings: The return of the king]. Translation by Agustin Otsoa. Tafalla: Txalaparta.

Trask, Robert L. 2003. The noun phrase: Nouns, determiners and modifiers; pronouns and names. In José Ignacio Hualde & Jon Ortiz de Urbina (eds.), *A grammar of Basque* (Mouton Grammar Library 26), 113–170. Berlin & New York: Mouton de Gruyter.

Van Langendonck, Willy. 2007. *Theory and typology of proper names* (Trends in linguistics. Studies and monographs 168). Berlin & New York: Mouton de Gruyter.

Zabala, Pello. 2000. *Naturaren mintzoa: Egunez egun, sasoien gurpilean* [Nature's speech: Day by day in the wheel of seasons]. Irun: Alberdania.

Zuazo, Koldo. 2008. *Euskalkiak: Euskararen dialektoak* [Euskalkiak: The dialects of Basque]. Donostia: Elkar.

Thomas Stolz and Nataliya Levkovych
On *Special Onymic Grammar (SOG)*: Definiteness markers in Fijian and selected Austronesian languages

Abstract: This study looks into the article system of Fijian languages to determine whether there is evidence of *Special Onymic Grammar (SOG)*. The basic distinction of proper names vs. common nouns has been acknowledged in the extant literature. The data are checked for evidence of the existence of a homogenous category of proper names or a split that separates personal names from place names. The focus is on morphosyntactic issues in Fijian and related languages. The findings are evaluated to the benefit of the research program dedicated to the morphosyntactic patterns of place names in cross-linguistic perspective.

Keywords: Fijian, morphosyntax, personal names, place names, Special Onymic Grammar.

1 Introduction

The ultimate goal of this paper is to lend support to the concept of *Special Onymic Grammar (SOG)* as coined by Nübling et al. (2015: 64). More specifically, we aim at proving empirically that there is evidence of the existence of more than one kind of SOG since, at least for certain languages, the members of the class of proper names can be shown to behave markedly different on the morphosyntactic level (Handschuh and Dammel 2019: 453–454). The differential behavior is particularly conspicuous in the case of place names whose grammatical properties frequently do not (fully) conform to those of personal names and at the same time also deviate from those of common nouns. This means that SOG can be further divided in two branches, namely *Special Anthroponymic Grammar (SAG)* and *Special Toponymic Grammar (STG)*. Aspects of SAG or STG

Thomas Stolz: Universität Bremen, FB10: Linguistik, Universitäts-Boulevard 13, D-28 359 Bremen, stolz@uni-bremen.de
Nataliya Levkovych: Universität Bremen, FB10: Linguistik, Universitäts-Boulevard 13, D-28 359 Bremen, levkov@uni-bremen.de

https://doi.org/10.1515/9783110672626-009

in individual languages have been treated in numerous publications such as Nübling (2017a) on phenomena in German. Comparative issues within genetically defined groups of languages are addressed, for instance, in Nübling (2017b) with reference to the Germanic branch of Indo-European. Within the wider framework of language typology which is relevant for this contribution, SAG is in the focus of a variety of studies among which we find Handschuh (2017, 2019) whereas STG is advocated for by Stolz et al. (2017a, 2017b, 2018). This study is a combination of the aforementioned approaches since it highlights SAG and STG not only with reference to an individual language (i.e. Fijian) but also takes account of data from genetically related languages (i.e. Austronesian) and interprets the findings from the point of view of language typology and the grammar of names in general. Since definiteness is crucial for the phenomena discussed below, this study fits in with several of the contributions of this edited volume in which the employment of the definite article with different classes of proper names is discussed (cf. the articles by Caro Reina, Helmbrecht, D'hulst et al., and Salaberri in this volume).

To this end, we proceed along the following lines. In Section 2, we turn our attention to the onymic markers and the role Fijian plays in the extant literature on this category. Section 3 presents the Fijian data and their analysis. Comparative aspects are addressed in Section 4. The conclusions are drawn in Section 5.

2 Fijian as a showcase of onymic markers

Our point of departure is defined by the ideal situation of a language which consistently distinguishes common nouns from proper names by way of overtly marking them for word-class membership. This scenario is sketched by Nübling et al. (2015: 64) as follows:

> Theoretisch könnte man das gesamte Onomastikon mit dem Lexikon gleichschalten und die gemeinte Kategorie jedes Mal bspw. mit einem E[igen]N[amen]- bzw. APP[ellativum]-spezifischen Präfix markieren.[1]

---

[1] Our translation: "Theoretically the entire onomasticon could be forced into line with the lexicon and the category at issue could be marked each time for instance, with a specific prefix for proper names or common nouns".

According to this division, the language in question would have two distinct morphological markers, namely one which identifies proper names as opposed to the other which marks common nouns.

Languages which come relatively close to this ideal abound in Oceania where they represent the East-Malayo-Polynesian branch of Austronesian. Handschuh (2018) surveys the Oceanic givens with special focus on personal names. More generally, Lynch et al. (2011: 38) describe a widely common feature of the same language group as follows:

> Many Oceanic languages have articles that precede a noun phrase. These often make a distinction between singular and plural, and between common and proper, and sometimes make a more fine-grained set of semantic contrasts than this. In Fijian, for example, the distinction between common and proper is marked by the preposed articles *na* and *o* respectively […]. Noun phrases with generic or locative/temporal reference generally do not appear with any article.

Not only in this quote does the Melanesian language Fijian serve the purpose of illustrating the formal distinction of proper names and common nouns. As Nübling (2005: 25–26) rightfully states, in the extant literature, the opposition of the proprial and common-noun markers in Fijian is mentioned time and again with reference to Hockett (1958). The frequency of citations of and quotes from this source notwithstanding, it makes sense to recapitulate what is said exactly in the original. There are two paragraphs in which Hockett (1958: 311–312) refers to the Fijian situation. We quote these passages verbatim, namely first:

> In Fijian, /mata/ 'day' is preceded by /na/ when it is the subject of a clause, but /viti/ 'Fiji' is preceded instead by /ko/. /na/ and /ko/ are two distinct particles, not different inflected forms of a single stem. Yet the choice of /na/ or /ko/ establishes a twofold classification of all Fijian nouns and noun phrases: names of specific people and places belong to the /ko/ class, common nouns to the /na/ class. Since a common noun or noun phrase is sometimes adopted as the name of a person or place, the classification is not quite mutually exclusive: /na vanua levu/ would mean '(a) big island', while /ko vanua levu/ is the name of a specific large island in the Fijian archipelago. (Hockett 1958: 230–231)

and later on in slightly different wording:

In Fijian, a word used as a proper of person or place is marked by the preceding particle /ko/, while words used as "ordinary" names of things are marked in the same syntactical circumstances by /na/: /na vanua levu/ 'the (*or* a) big land, big island' but /ko vanua levu/ 'Big Island' as the name of the largest island of the Fiji group. (Hockett 1958: 311–312)

The author employs the same examples in both paragraphs. The different functions of *na* and *ko* are illustrated with complex NPs involving a postnominal adjectival attribute *levu* 'big' which modifies the noun *vanua* 'place, land'. In the earlier quote, Hockett (1958) mentions the subject function of the NP so that the question arises whether the NPs and the accompanying markers behave differently in a clause if they fulfill a function other than that of subject. Note also that the NPs are associated with definiteness and alternatively with indefiniteness since Hockett uses English *the* and *a* in his translations. Furthermore, Hockett (1958) puts the label particle *to* the markers under review whereas Lynch et al. (2011) use the term article and oppose *na* to *o* in lieu of *ko*. Further arguments concerning the status of the supposed onymic article are put forward in Reid (2002: 296–297) and Himmelmann (2005: 246). For the time being, we will use the neutral term markers.

Seiler (1986: 136) avoids taking sides as to this terminological issue when he states that "[f]or Fijian, C.F. Hockett reports the existence of two elements, *na* and *ko*, distinguishing between the class of common nouns from the class of names of specific people and places [...]".

This author also assumes a binary system proprial vs. common but the markers are not exactly the same as those mentioned by Lynch et al. (2011) in the previous quote. Nübling (2005: 26) bases her argumentation on the same paragraphs in Hockett's work but inadvertently mixes up the two markers (which she otherwise considers to be classifiers) so that she speaks of the "Voranstellung der onymischen Partikel *na* im Fidschi"[2] (Nübling 2005: 41) although her source of information, Hockett (1958: 311–312), identifies *ko* as onymic marker instead. Caro Reina (2014: 178) discusses the Fijian data independently of Hockett. With reference to Schütz (1985: 320) and Aranovich (2013) he postulates a minimal paradigm of two markers, namely *na* for common nouns and *o* for proper names (including place names). The author emphasizes that the fact that the proprial marker *o* is also used in combination with personal pronouns casts doubt on its exclusive status as onymic marker because it could turn out to be a case of an animacy-related marker. Again with reference

---

[2] Our translation: "preposing of the onymic particle *na* in Fijian".

to Hockett's classic text, Heidenkummer and Helmbrecht (2017: 16) mention the Fijian case and the *na-ko* paradigm of markers. The authors specifically argue that Fijian is different from Hoocąk because, in the Austronesian language, the proprial marker *ko* is also employed with place names.

There is thus a general consensus as to the binary structure of the paradigm of markers in Fijian. Moreover, the above authors agree that the categories distinguished by the markers are those of common nouns and proper names. It is uncontroversial that *na* functions as marker of common nouns whereas there is some variation on the side of the proprial marker which is given either as *ko* or *o*. Moreover, it is also commonly accepted that the functional domain of the proprial marker covers both personal names and place names.

In what follows, we demonstrate that the situation is not as straightforward in Fijian so that the picture of the marker system calls for being repainted at least partly. We argue that, if one aims at determining the role SAG and/or STG play in a given language system, it is not sufficient to analyze the data exclusively on the basis of the inventory of markers. It is true that we know that personal names and place names share the same marker. However, this does not mean that we know whether the marker has the same distribution for both classes of proper names. As the typology of onymic marking strategies provided by Nübling (2005) suggests, proper names may stand out in many different ways from the bulk of the nouny word-classes of a given language. More often than not, there are no dedicated proprial markers in the first place but more or less subtle differences such as combinatory restrictions, etc. which show that proper names obey rules which are at least slightly different from those of the common nouns. The differences are thus often covert since they only surface in the interaction with other syntactic units. We claim that what counts for the differentiation of proper names and common nouns also holds for the possibility of distinguishing place names from personal names. This is what Section 3 is supposed to prove.

3 Fijian data explored

3.1 Proper names and common nouns in NPs

Caro Reina (2014: 178) presents a set of Fijian examples which consist of syntactically isolated NPs to show that proper names take a different marker from common nouns but at the same time share this property with personal pronouns. We adapt his examples in (1)–(2) below.[3]

(1) common noun
na *vale*
COMM house
'**a/the** house'

(2) proper names and pronouns

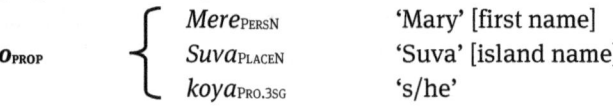

The four NPs in (1)–(2) invite an interpretation according to which there is a formal divide with common nouns on one side and proper names and pronouns on the other. Superficially, this interpretation seems to be fully corroborated if we take account of sentential examples of the use of comparable NPs in Fijian. In (3)–(6), common nouns, proper names, and personal pronouns are embedded in a syntactic context.

(3) Common noun (Schütz and Komaitai 1971: 3)
E vei beka [na mākete]$_{NP}$?
LOC where please [COMM **market**]$_{SUBJECT}$
'Where is **[the market]**, please?'

(4) Personal name (Schütz and Komaitai 1971: 31)
E vei [o Niko]$_{NP}$?
LOC where [PROP **Nico**]$_{SUBJECT}$
'Where is **[Nico]**?'

(5) Place name (Schütz and Komaitai 1971: 85)
[Ko Viti]$_{NP}$ e koto e
[PROP **Fiji**]$_{SUBJECT}$ TMA lie LOC
na ceva i rā ni wasa Pasifika.
COMM south DIR down of sea Pacific
'**[Fiji]** is situated down south in the Pacific Ocean.'

---

3 We mostly respect the glosses and English translations of the sources from which we take the examples. Additional square brackets and indexes are ours. Some glosses have been modified to reduce variation. Boldface marks out those parts of an example which are focused upon in the ensuing discussion. In sentential examples, we use additional indexed square brackets.

(6) Personal pronoun (Schütz and Komaitai 1971: 13)
E	dua	na	tūraga	[o	koya]$_{NP}$.
be	INDEF	COMM	man	[PROP	s/he]$_{SUBJECT}$

'[He] is a man.'

The differential behavior of common nouns and proper names is directly visible from the comparison of the NP *na mākete* 'the market' in (3) with the bracketed NPs in (4)–(6). In all other regards, the examples (4)–(6) do not speak for themselves because they raise several issues which need to be commented upon.[4]

What strikes the eye immediately is the use of two supposedly different markers with personal names and place names. In (4), the personal name *Niko* is preceded by *o* whereas in (5), the place name *Viti* is accompanied by *ko*. However, what superficially appears to be a formal distinction of two names classes turns out to be a matter of register. The proprial marker *o* is typical of spoken Fijian and thus abounds in those parts of the source we have consulted where the data are presented in dialogues. In contrast, example (5) has a certain bookish flair about it. This comes as no surprise because *ko* is mainly used in written Fijian or formal speech. Example (5) is taken from a written document that describes the islands of Fiji to foreigners. That *ko* can be replaced with *o* in a similar context is shown in (7).

(7) Place name (Schütz and Komaitai 1971: 179)
[O	Rotuma]$_{NP}$	e	koto		
[PROP	Rotuma]$_{SUBJECT}$	TMA	lie		
e	na	vualiku	kei	Vanua Levu.	
LOC	COMM	north	with	Vanua Levu	

'[Rotuma] lies to the north of Vanua Levu.'

Thus, the co-existence of *ko* and *o* constitutes a case of stylistically motivated variation of otherwise functionally equivalent allomorphs (Schütz 1985: 320–321).[5] The disagreement among the authors mentioned in the foregoing section

---

[4] For reasons of space, we skip discussing the proprial interrogative *o cei* (~ *i cei*) 'who? whatplace?' which involves the proprial marker o as opposed to the common interrogative *(n)a cava* 'what?' which comprises the common-noun marker *na* (Schütz 2014: 196, 226).

[5] Schütz (1985: 41) argues that the alternation of *o* ~ *ko* in present-day Fijian is result of standardization efforts by missionaries in the 19th century who mixed the properties of different dialects to create a common Fijian variety only to impose a rule according to which *o* should be used in sentence-initial position whereas *ko* is used in mid-sentence – a rule that is ridiculed by Schütz (1985: 41) as nonsensical (cf. example (5) with initial *ko* vs. examples (4) and (6) with sentence-medial *o*).

as to the outer form of the proprial marker is thus easily explained: they simply refer to two different registers of Fijian.

The second phenomenon which is of interest is the use of the proprial marker *o* in combination with the pronoun of the 3rd person singular *koya* in (6). As mentioned above already, Caro Reina (2014) considers cases like *o koya* to be, at least potentially, pieces of counterevidence which speak against an interpretation of *o* as a proprial marker. If *o* is re-interpreted as a marker of extended animacy, then it would only be logical that the putative common-noun marker *na* turns into the marker of inanimacy. However, *na* is used with definite NPs of different animacy. A third form can be used with indefinite NPs. It makes sense to consider this an example of Differential Definiteness Marking.[6] It is clear that *o koya* is related to the demonstrative *((k)o)yā* 'that' which Schütz (2014: 242–243) describes as referring to location near the 3rd person, i.e. as a distal demonstrative. The short form is common especially in fast speech (Schütz 2014: 245). As demonstrative *((k)o)yā* 'that' is insensitive to animacy, meaning: it can also be used for inanimate referents as in (8).

(8)   Demonstrative (Schütz 2014: 245)
      [Oyā]$_{NP}$ na   no-qu   koro.
      [DEM.DIS]$_{SUBJECT}$ COMM   PC-1SG   village
      '[That]'s my village.'

According to Schütz (2014: 198–199), there is a class of morphologically complex proper pronouns for all persons and numbers whose initial component is *o* as shown in Table 1.

Table 1: Fijian proper pronouns (Schütz 2014: 199).

Number	1st EXCL	1st INCL	2nd	3rd
SG	*o yau*	---	*o iko*	*o koya*
DU	*o keirau*	*o kēdaru*	*o kemudrau*	*o rau*
TR	*o keitou*	*o kedatou*	*o kemudou*	*o iratou*
PL	*o keimami*	*o keda*	*o kemunī*	*o ira*

---

[6] We gratefully acknowledge that these ideas are based on suggestions made by one of our anonymous reviewers.

The function of the proper pronouns is that of specifying a subject in the sense that they are used in addition to a co-present subject be it pronominal or lexical. Schütz (2014: 198–199) claims that the proper pronouns "do not add any semantic information, but their function could be vaguely labeled as emphasis". If the subject of a given sentence is pronominal, the proper pronouns are used in addition to the subject pronouns in Table 2. Note that the *o* in the column of the 2nd person is not related to the proprial marker. Bracketed segments are optional in spoken Fijian (Schütz 2014: 33).

Table 2: Fijian subject pronouns (Schütz 2014: 28).

Number	1st EXCL	1st INCL	2nd	3rd
SG	au	---	o	e
DU	keirau	(e)daru	(o)drau	(e)rau
TR	keitou	((e)da)tou	(o)dou	(e)ratou
PL	keimami	(e)da	(o)nī	(e)ra

Subject pronouns and proper pronouns typically co-occur under topicalization as in (9).

(9) Topicalization (Schütz 2014: 34)
   [O   yau]~NP~, *au*   *sā*   *lako*
   [PROP 1SG]~TOP~   1SG~SUBJECT~ TMA   go
   '[As for me], I go.'

The use of the proprial pronouns thus is motivated by pragmatics. Interestingly, Schütz (2014: 30) argues against the hypothesis "that nonsingular third person subjects refer only (or nearly so) to humans. Data do not support this hypothesis. The criterion for nonsingular subjects seems to be not humanness, but *individuality*" [original italics].

For the proper pronouns, this means that they are licit also in contexts in which they are co-referential with a subject pronoun that has an inanimate referent as in (10).

(10) Topicalization/inanimate referent (Schütz 2014: 199)[7]
[O   iratou   oqori]$_{NP}$,   *eratou*   *raba-i-le~levu.*
[PROP 3TR   DEM.MID]$_{TOP}$ 3TR$_{SUBJECT}$ wide-INS-RED~stout
'**[As for those three there]**, they are wide.'

The pronouns refer to different types of *masi* 'tapa', i.e. to inanimate objects mentioned in the previous context. It is therefore not necessarily the case that the presence of *o* is triggered by animacy. Subjecthood and pragmatic prominence are the crucial factors. This is also the case with proper names since they require the presence of the proprial marker when they function as subjects as in (11) or topicalized constituent as in (12).

(11) Personal name as subject (Schütz 2014: 198)
Sā   vaka-yakavi   tiko   [o   Jone]$_{NP}$
TMA   MAN-evening   CNT   [PROP Jone]$_{SUBJECT}$
'**[Jone]** is having supper.'

(12) Personal name as topic (Schütz 2014: 198)
[O   Jone]$_{NP}$,   *au   rai-c-a*
[PROP JONE]$_{TOP}$ 1SG$_{SUBJECT}$ see-TRANS-3SG
'**[As for Jone]**, I saw him.'

In (12), the topicalization affects the object of the matrix clause. However, the proprial marker can only be used with a proper name in object function when the NP is fronted and functions as extra-sentential topic. If a personal name is the regular object, the proprial marker is blocked as in (13). The absence of the proprial marker is indicated by the symbol Ø.

(13) Personal name as internal argument (Schütz 2014: 198)
E   ā   rai-ci   [Ø   Paula]$_{NP}$
3SG$_{SUBJECT}$ PT   see-TRANS   [Ø   Paula]$_{OBJECT}$
'S/he saw **[Paula]**.'

Parallel examples of place names being used in object function are not easy to come by, but there is at least (14).[8]

---

[7] Schütz (2014: 199) translates 'Those (just listed) are wide'. His glosses are incomplete. We have added the missing parts.

[8] Further examples can be found in Alderete (1998: 20–21).

(14) Place name as internal argument (van Urk 2020: 324)
au   *rai-ci*   *[Ø*   *Viti]*$_{NP}$
1SG$_{SUBJECT}$ see-TRANS   [Ø   Fiji]$_{OBJECT}$
'I saw **[Fiji]**.'

Both classes of proper names behave the same insofar as the proprial marker is present/absent under identical conditions: external arguments and topics require the presence of *o* whereas the use of *o* is ungrammatical for proper names as internal arguments. There is thus a rule-based alternation of *o* and Ø (van Urk 2020).

In contrast, common nouns always keep the common-noun marker as shown in (15)–(16) (and (3) above).

(15) Common noun as internal argument (Schütz 1985: 322)
āū   *ā*   *caqe-t-a*   *[na*   *polo]*$_{NP}$.
1SG   PT   kick-TRANS-3SG   [COMM   ball]$_{OBJECT}$
'I kicked **[the ball]**.'

(16) Common noun$_{OBJECT}$ as topic (Schütz 1985: 322)
[na   *polo]*$_{NP}$ *āū*   *ā*   *caqe-t-a.*
[COMM   ball]$_{TOPIC}$ 1SG   PT   kick-TRANS-3SG
'**[As for the ball]**, I kicked it.'

Does this mean that the common-noun marker is of a different kind as compared to that of the proper names? Do the markers fulfill mutually incompatible functions?

As to the function of the above markers, Lynch (1998: 110–111) assumes that "*[n]a* is the **common article** [original boldface] and is used before other nouns that are definite in some sense. Indefinite nouns […] are not marked by articles". Schütz (2014: 199, footnote 6) agrees with this opinion arguing that definiteness is an issue only with common nouns whereas a "proper NP […] is already semantically definite in that it refers to a specific person or place". On the basis of this argumentation, it can be assumed that the common-noun marker and the proper marker belong to two different functional domains. If we look back to example (6), we notice, however, that the rule postulated by Lynch in the above quote does not capture all instances of indefiniteness. The NP *dua*$_{INDEF}$ *na*$_{COMM}$ *tūraga* 'a man' hosts the leftmost constituent *dua* 'one' which Schütz (2014: 204–205) classifies as marker of indefiniteness. There is thus a paradox since the indefiniteness marker *dua* co-occurs with the supposed definiteness marker *na*. Bare common nouns which Lynch analyzes as indefinite NPs have generic reference. When they are attested VP-internally, they can be understood as instances of incorporation (Aranovich 2013), not the least because the verb is not marked for transitivity as in (17)–(18).

(17) Referential internal argument (Lynch 1998: 111)
 E gunu-**va** **na** yaqona o Seru.
 3SG drink-TRANS COMM kava PROP Seru
 'Seru is drinking **the** kava.'
(18) Incorporation (Lynch 1998: 111)[9]
 E gunu yaqona o Seru.
 3SG drink kava PROP Seru
 'Seru is drinking kava.'

The above differences between proper names and common nouns as internal arguments are rubricated as instances of a special kind of Differential Object Marking (DOM) in the extant literature (Alderete 1998). Whatever the optimal approach to the problem might be, the empirical data are indicative of pronounced structural differences between proper names and common nouns in the morphosyntactic domain.

What we have seen so far suggests that there is a homogeneous class of proper names covering both personal names and place names. In Section 3.2, we provide examples of the differential behavior of the two kinds of names so that it makes sense to assume a binary split in two grammatically relevant subclasses of proper names.

3.2 Proper names and common nouns in PPs

When Fijian NPs are complements of prepositions, they display structural heterogeneity (Geraghty 1976: 507, 520). The differences between common nouns, personal names, and places names are especially discernible in examples like (19)–(21).[10]

---

9 Original English translation (we would have preferred 'Seru is kava-drinking/Seru is a kava-drinker'), glosses modified by us.
10 In the examples presented in this section, we index the PPs for the spatial relation they express.

(19) Common noun (Schütz and Komaitai 1971: 25)
      E   cakacaka tiko   *[mai   [na   koro]_NP]_PP*
      3SG   work   CNT   [ABL   [COMM   village]]_PLACE
      ni   vuli   na   tamata oqō.
      for   learn   COMM   people here
      'He is working **[in (lit. from) [the village]]** to teach the people here.'

(20) Personal name (Schütz and Komaitai 1971: 96)
      Mo   kacivi   rau   *[mai   [o   Samu   kei   Viliame]_NP]_PP*.
      SUB:2SG   call:PASS   3DU   [ABL   [PROP   Sam   and   William]]_SOURCE
      'You should be called **[by [Sam and William]]**.'

(21) Place name (Schütz and Komaitai 1971: 5)[11]
      Au   *[mai   [Ø   Awai]_NP]_PP*.
      1SG   [ABL   [Ø   Hawaii]]_SOURCE
      'I come **[from Hawaii]**.'

As shown in (19)–(20), the (multifunctional) ablative preposition *mai* 'from' takes common nouns and personal names together with their markers as complements. Example (20) is highly marked and will receive further attention below. The preposition *mai* and place names interact differently in the sense that the latter are never accompanied by the proprial marker – and this rule holds for combinations with every preposition of Fijian. Schütz (2014: 216, 224) contrasts e_{PREP} na_{COMM} *vale-ni-kana* 'in the restaurant' with e_{PREP} Ø *Suva* 'in/at Suva' where the locative preposition *e* 'in, at' takes a common noun vs. a place name as complement. The common noun can fulfill this function only if the common-noun marker *na* is present, the place name *Suva*, however, is directly governed by the preposition. Sentential examples are provided in (22)–(23).

(22) Common noun (Geraghty 1976: 508)
      E   tiko   *[e   [na   koro]_NP]_PP*
      3SG   stay   [LOC   [COMM   village]]_PLACE
      'He is **[in the village]**.'

(23) Place name (Geraghty 1976: 508)
      Au   tiko   *[e   [Ø   Nansori]_NP]_PP*
      1SG   stay   [LOC   [Ø   Nansori]]_PLACE
      'I am **[in Nansori]**.'

---

[11] For interesting parallel cases in Romance we refer the readers to Caro Reina (2020 and this volume).

The directional preposition *ki (~ i)*[12] 'to, at' makes similar requirements for its complements as shown in (24)–(25).

(24) Common noun (Schütz and Komaitai 1971: 25)
    E   lako   [ki   [na   bose]_{NP}]_{PP}   o   koya
    3SG go   [DIR [COMM meeting]]_{GOAL} PROP s/he
    'He goes **[to the meeting]**.'

(25) Place name (Schütz and Komaitai 1971: 25)
    E   lako   [ki   [Ø   Suva]_{NP}]_{PP}   o   koya
    3SG go   [DIR [Ø   Suva]]_{GOAL}   PROP s/he
    'He goes **[to Suva]**.'

Place names are distinguished from common nouns by the mandatory presence of the common-noun marker with the latter whereas no marker is used with the place names.

Personal names are again a different story as they are excluded from PPs headed by either *e* or *ki (~ i)*. Personal names and personal pronouns possess an all-purpose spatial preposition of their own which is *vei* 'from, for, to' which Schütz (2014: 220) glosses as ablative. The examples (26)–(27) illustrate the use of this preposition.

(26) *Vei* with personal name (Schütz 2014: 220)[13]
    E   ā   vaka-tabataba gā   [vei   [Ø   Mārica]_{NP}]_{PP}. ko   Leone
    3SG PT CAUS-silence   LIM   [ABL [Ø   Mārica]]_{GOAL}   PROP Leone
    'Leone signaled silence **[to Mārica]**.'

(27) *Vei* with personal pronoun (Schütz 2014: 221)[14]
    O   iko   m-o   tuku-n-a   [vei Ø   [ira]_{NP}]_{PP}.
    PROP 2SG SUB-2SG tell-TRANS-3SG [ABL Ø   [3PL]]_{GOAL}
    'You tell **[them]**.'

Schütz (2014: 220) remarks that place names do not follow personal names in this case because place names "occur in *e*, *mai*, and *i* phrases". As to the directional prepositions *mai* and *ki (~ i)*, Schütz (2014: 218) states that "place names pattern not with propers, but with commons". These statements are perhaps

---

12 Whether *ki* and *i* are allomorphs of each other is a hotly debated issue (Schütz 2014: 224) the solution of which is, however, of no relevance for this study.
13 Schütz (2014: 220) puts English *to* in brackets.
14 Schütz (2014: 221) includes information on the number categories in the translation.

overly suggestive. The claim that place names pattern with common nouns does not entail that the PPs have identical internal structure. PPs with a common noun as complement count three constituents whereas those with a place name as complement have a binary structure because the proper marker is generally absent from the PP. What common nouns and place names share is their compatibility with the same spatial prepositions to the exclusion of personal names. In contrast, place names and personal names cannot combine with the same prepositions but they behave similarly to each other insofar as they come in the shape of bare nouns (i.e. without proprial marker) when they are complements. The similarities and dissimilarities in the interaction of the nouny word-classes with spatial prepositions is summarized in Table 3. For similar tables with larger inventories of prepositions, the reader is referred to Geraghty (1976: 519) and Dixon (1988: 152). Grey shading marks out identically filled cells in the same row.

Table 3: Spatial relations and definiteness markers with proper names and common nouns in Fijian.

Spatial relation	Personal names	Place names	Common nouns
LOC	vei	e	e
ABL	vei	mai	mai
DIR	vei	(k)i	(k)i
MARKER	∅	∅	na

Personal names and common nouns have maximally different profiles whereas place names share properties with both. There is thus not only tangible evidence for SOG but also for SAG and STG since it is not possible to capture their morphosyntactic behavior with one and the same rule.

Before we close this section, a comment on example (20) is called for. This example is special for two reasons. According to Table 3, personal names are excluded from PPs headed by the ablative preposition *mai*. This restriction is violated against because in (20) *mai* takes two coordinated personal names as complement. Secondly, the leftmost of these coordinated personal names is accompanied by the proper marker *o*. The presence of the marker is problematic in the sense that, in PPs, proper names are never accompanied by their marker. On account of these illicit properties, one might question the status of (20) as a specimen of correct or native Fijian. In point of fact, Schütz and Nawadara (1972) rebut the idea that there is a passive voice in Fijian, in the first place.

Geraghty (1976: 519) observes that in certain styles "translationese" takes place and is visible in "the use of *mai* and *maivei* to translate the agent of the English passive construction. [...] Again, this usage is not found in informal speech, but is perhaps the salient feature of Radio Fiji news, where a quick translation of the English passive is often needed".

It is therefore conceivable that example (20) represents a grammatical copy from English.[15] The English pattern from which the construction is copied does not explain why the proprial marker is retained. If we are not facing a simple typographical error, the presence of *o* in (20) might be explicable along the following lines. The proprial marker is associated most strongly with subjecthood and/or topicality. Since syntactic subjects often correspond to the semantic role agent, chances are that the agent role of *Samu kei Viliame*[16] has triggered an association with the subject status so that the use of the proprial marker seemed to be justified. This merely speculative interpretation does in no way affect the conclusion that Fijian gives evidence of SAG and STG albeit in a subtle way. There are no distinctive markers for place names as opposed to personal names and *vice versa*. Nevertheless, the grammars of the two sub-classes of proper names are not identical, i.e. the use of the same proprial marker does not imply that all proper names behave the same morphosyntactically.

In the subsequent section, we cursorily take account of comparative data to see how the Fijian findings fit into the general picture.

4 Comparative data

For the purpose of comparison, we focus on phenomena which occur in the NP. Only for Maori do we also look at the PP. The reason for this practice is the scarcity of reliable empirical data which were accessible to us at the time of writing this paper.

---

[15] On the basis of parallels in Romance languages where native names lose the article whereas foreign names keep it, an anonymous reviewer suggests that we are dealing with the effect of foreignness also in Fiji.

[16] The doubtful status of the example notwithstanding, the coordinated NPs behave exactly as predicted by Aranovich (2013: 474–475) because the marker on the second NP falls victim to Equi-Deletion. Note that Equi-Deletion does not affect the common-noun marker.

4.1 Nadrogā

Nadrogā is a Western Fijian language spoken in the southwest of Viti Levu, i.e. it is co-territorial and genetically affiliated with Fijian. There are three distinct markers, namely *na* for common nouns, *o* for personal names as well as personal pronouns and kin terms, and *i* for place names. The examples (28)–(30) illustrate the use of these markers.

(28) Common noun (Geraghty 2011: 841)
 Qu tolavia [na lewa]$_{NP}$.
 1SG see:TRANS:3SG [COMM woman]$_{OBJECT}$
 'I saw **[the woman]**.'
(29) Personal name/kin term (Geraghty 2011: 837)
 Eri vica [o luve-muru]$_{NP}$?
 3PAC how_many [PROP$_{PERSON}$ child-2DU.POSS]$_{OBJECT}$
 'How many **[children]** do you two have?'
(30) Place name (Geraghty 2011: 837)
 Qī matā tolavia [i Huva]$_{NP}$.
 1SG want see:TRANS:3SG [PROP$_{PLACE}$ Suva]$_{OBJECT}$
 'I want to see **[Suva]**.'

In contrast to the Fijian case, there are not only three markers, two of which formally distinguish personal names from place names, but Nadrogā also allows for NPs of all classes to keep their markers also in object function (as in (30)). As far as we can judge from the relatively limited documentation of the language, zero-marking is not an established strategy.

In Fijian, one has to dig deep, in a manner of speaking, to identify formal morphosyntactic differences between the two sub-classes of proper names. In Nadrogā, on the other hand, personal names and place names go their separate ways already on the level of the markers. The evidence for SAG and STG is thus almost directly visible.

4.2 Maori

The situation in the Polynesian language Maori (New Zealand) resembles the Fijian case in the sense that there is a heavy dose of covertness involved. There are two markers, viz. proprial *a* and common *te*. With the proviso that "the rules governing [...] the use [of the articles] are not clear in all cases", Bauer (1993:

109) states that the use of markers with Maori nouns is obligatory except in the following cases:
- "before local nouns including place names;"
- "however, when these function as sentence subject, they take the personal article *a*,"
- "pronouns and personal names do not require an article following prepositions other than *i, ki, kei, hei*."
- In addition, the topic marker *ko* excludes the use of the definiteness marker *a* – a rule not mentioned in Bauer's description.[17]

The other two exceptions to the general use of the markers are more interesting for the topic at hand. Place names are reported not to take an article unless they function as subject-NP of a clause. If they have this function, they are accompanied by proprial *a* which is also the marker of personal names. Personal names on the other hand may do without the article *a* in PPs which are headed by prepositions other than those enumerated above. Common nouns (other than relational "local nouns" like *runga* 'top') are always preceded by a marker (which usually comes in the shape of *te*).

Accordingly, there are two sub-domains in which personal names and place names display identical morphosyntactic behavior. As subject-NPs both of them require the presence of the proprial marker, and as complements of certain prepositions, neither personal names nor place names combine with any marker.

This parallel behavior does not exhaust all syntactic contexts though. Personal names and place names differ from each other insofar as place names do not take a marker in any context except that of the subject whereas personal names also need the marker to fulfill syntactic functions beyond that of the subject.

Furthermore, the proprial marker is obligatory with personal names also with a small set of four prepositions which, however, do not trigger the use of the article with a place name as complement. The crucial similarities and dissimilarities of the nouny word-classes are illustrated by the examples in (31)–(34).

(31) Common noun (Bauer 1993: 93)
Kei te   iri   [te   whakaahua]$_{NP}$
TMA   hang [COMM   picture]$_{SUBJECT}$
[i   runga]$_{PP}$ [i   [te   pakitara]$_{NP}$]$_{PP}$
[LOC   top]   [LOC   [COMM   wall]]
'[The picture] is hanging [on [the wall]].'

---

17 We are grateful to one of our anonymous reviewers for advising us in this matter.

(32) Personal name – subject and PP-complement (Bauer 1993: 92)
Ka   aahua   pukuriri [a   Tamahae]$_{NP}$ [ki   [a   Rewi]]$_{NP}$$_{PP}$
TMA   somewhat   angry   [PROP   Tamahae]$_{SUBJECT}$ [DIR [PROP Rewi]]
'[Tamahae] was somewhat angry [with Rewi].'

(33) Place name – subject (Bauer 1993: 329)
[A   Te Kao]$_{NP}$ kei teeraa taha   o   Kaitaaia.
[PROP   Te Kao]$_{SUBJECT}$ at that   side   GEN   Kaitaia
'[Te Kao] is beyond Kaitaia.'

(34) Place name – PP-complement (Bauer 1993: 129)
I   haere   atu   a   ia   [ki   [Ingarangi me Amerika]$_{NP}$]$_{PP}$.
TMA   move   away   PROP 3SG   [DIR   [England and America]]
'He travelled [to England and America].'

In (32), the personal name *Rewi* is accompanied by the proprial marker *a* because the governing preposition is *ki* which requires the presence of the marker if the complement is a personal name. As shown in (34), place names which are the complement of the same preposition *ki* do not take the proprial marker. Common nouns are always accompanied by *te* (as in (31)) whereas the presence of *a* is compulsory for both subclasses of proper names only in subject function (as in (32)–(33)). Table 4 provides a synopsis of the distribution patterns of the three nouny word classes. Similar behavior in a given row is highlighted in grey.

Table 4: Presence and absence of markers in Maori.

Criterion	Place names	Personal names	Common nouns
subject	a	a	te
[(C)X-i_{PREP} __]$_{PP}$	Ø	a	te
[Y$_{PREP}$ __]$_{PP}$	Ø	Ø	te

The proof of the existence of SOG in Maori is straightforward since there is the opposition of the common-noun marker *te* and the proprial marker *a*. As to SAG and STG, the evidence is less direct because it can only be gained from a distributional analysis of the contexts in which the proprial marker is mandatory or blocked. This analysis shows that personal names and place names do not constitute a homogeneous class in Maori.

4.3 Taiof

Taiof is an Austronesian language spoken on the homonymous island in the vicinity of Bougainville. The language boasts four so-called genders, namely common nouns I and II, personal names, and place names (Ross 2011: 432). Common nouns take the marker *a* or *i* in the singular whereas personal names take the marker *e*. In contrast, place names are never accompanied by any marker. Examples are presented in (35)–(37).

(35) Common noun (Ross 2011: 435)
 [A   numa avai]_NP=na   foun.
 [COMM house DEM.PROX]_SUBJECT=COMM new
 '[This house] is new.'

(35) Personal name (Ross 2011: 437)
 [E   Maras]_NP   o   keta=no-n   to-nai.
 [PROP_PERSON Maras]_SUBJECT REAL speak=IMPF-3SG PREP-3SG
 '[Maras] is talking to him.'

(36) Place name (Ross 2011: 435)
 [Ø   Kieta]_NP   i   vivian.
 [Ø   Kieta]_SUBJECT   PREP far_away
 '[Kieta (town)] is far away.'

Taiof provides evidence of a tripartite system which differs from that of Nadrogā because place names are exempt completely from marking noun-class membership. Taiof place names are zero-marked throughout the grammatical system of the language. The fact that proper names do not take the same markers as common nouns suffices as proof of SOG. SAG and STG manifest themselves in the opposition of overt marking for personal names and zero-marking of place names.

4.4 Chamorro

We conclude the section on comparative cases with a glance at Chamorro, an internal isolate spoken in the Marianas (Micronesia). This language has a ternary system of nominal word-classes. Common nouns take the marker *i* (focus)/*ni* ~ *nu* (non-focus). For personal names, the marker *si* (focus)/*as* (non-focus) is used (Topping and Dungca 1973: 245–247). Place names are accompanied by the marker *iya* which is neutral as to the distinction of focus vs. non-focus.[18] In (37)–(39), the use of these markers is illustrated.

(37) Common noun (Topping and Dungca 1973: 132)
[I   patgon]ₙₚ   ha   hatsa   [i   lamasa]ₙₚ
[COMM   child]ₐcₜₒᵣ   3SG.ERG   lift   [COMM   table]ᵤₙ𝒹ₑᵣ𝒢ₒₑᵣ
'[The child] lifted [the table].'

(38) Personal name (Topping and Dungca 1973: 131)
Chiniku   [si   Maria]ₙₚ   [as   Juan]ₙₚ
<PASS>kiss   [PROP_PERSON   Maria]_FOCUS   [PROP_PERSON   Juan]_NON_FOCUS
'[Mary] was kissed [by Juan].'

(39) Place name (Topping and Dungca 1973: 131)
[Iya   Umatac]ₙₚ   dikike'   na   songsong
[PROP_PLACE   Umatac]_FOCUS   small   LINK   village
'[Umatac] is a small town.'

The data almost speak for themselves. Chamorro resembles Nadrogā and Taiof in the sense that it provides tangible proof of SOG already in the marker system but also because personal names and place names are treated differently in the grammar. Chamorro attests to both SAG and STG the distinction being directly visible from the set of markers that is employed.

5 Conclusions

The five Austronesian languages touched upon in this study yield interesting results in relation to SOG, SAG, and STG. The distinction of personal names vs. place names interacts with further factors such as focus, case, phrase type, and preposition type. The patterns we have identified in the previous sections are summarized in Table 5. Grey shaded cells host identical markers in a given row.

---

18 An in-depth study of the Chamorro place-name marker *iya* is provided in Stolz (2020).

Table 5: Marker systems of five Austronesian languages in NPs.

Language	Common nouns	Personal names	Place names
Fijian	na	(k)o	(k)o
Maori	te	a	a
Nadrogā	na	o	i
Chamorro	i / ni ~ nu	si / as	iya
Taiof	a / i	e	Ø

Only for Fijian and Maori is the identification of SAG and STG unnecessary because there are no distinct markers for the two subclasses of proper names. In the remaining three languages, it is sufficient to look at the markers to see that SAG and STG are at work. On this basis it is possible to put forward a hypothetical generalization over Oceanic languages according to which the existence of a place-name marker (including cases of zero-marking) distinct from a marker for common nouns goes along with the existence of a distinct personal-name marker as shown in Figure 1.

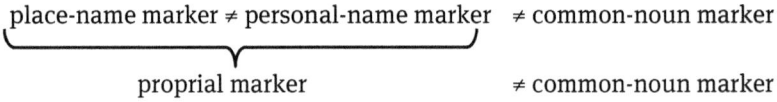

Figure 1: Hypothetical patterns (for Oceanic only).

For the time being, the scope of this generalization is limited to the Oceanic languages. Its validity has to be tested further empirically. In case it is corroborated by the evidence from further languages of this branch of Austronesia, one might think of universalizing the claim and check its tenability on the basis of a larger cross-linguistic sample. In the light of what we know already about the proprial markers in Hoocąk and regional varieties Catalan which are restricted to personal names (Heidenkummer and Helmbrecht 2017; Caro Reina 2014) and similar cases as reviewed by Nübling (2005) it seems questionable that the above patterns stands the test outside the Austronesian context. Its potential failure, however, raises the general question whether SOG can be shown to preferably follow certain paths to the detriment of others.

In this study, we have inspected the co-occurrence of definiteness markers and proper names in selected Austronesian languages with special focus on

Fijian in order to prove that there are enough pieces of evidence in support of the hypothesis that languages for which the existence of SOG has been established are also candidates for the more fine-grained distinction of SAG and STG. Fijian belongs to those languages in which the SAG-STG opposition is not obvious on superficial inspection. The grammatically relevant differences between personal names and places names come to the fore only in the interaction of these categories with other constituents in syntax. The comparative data from Maori strongly suggest that we are dealing with a relatively wide-spread phenomenon among Oceanic languages.

This phenomenon can be directly associated with Whorf's (1945: 5) concept of "cryptotype", i.e. with grammatical categories which lack dedicated markers but can be identified on the basis of combinatorial properties of the units at issue. Both in Fijian and Maori, the distinction of personal names and place names is cryptotypical in the sense that the marker is the same for both but
- in Maori, it reacts differently to its morphosyntactic contexts depending which of the two proprial sub-classes is the complement of a preposition, or
- in Fijian, different prepositions have to be used to express identical spatial relations with personal names and place names.

We argue that similar scenarios can be found all over the place not only within the Austronesian language family but also far beyond. This hypothesis deserves to be investigated in-depth in future studies.

As three of the languages reviewed in Section 4 show, the cryptotype is not the only realization form of the SAG-STG opposition. There are languages which distinguish the above categories formally already on the level of the markers. We therefore assume that the languages of the world can be divided in several classes, namely those without SOG vs. those with SOG whereas the latter can be subdivided further in languages with a binary proper-common distinction and those with SAG/STG. In the latter case, it is possible to distinguish languages whose systems operate on the basis of the cryptotype from those which employ overt dedicated marking strategies. Moreover, it is necessary to take stock of all those instances in which only one of the sub-classes of proper names attests to SOG. Which languages realize the pattern personal names vs. [common nouns + place names] and which ones witness place names vs. [common nouns + personal names]? How these types are distributed cross-linguistically and whether the proposed classification is linguistically meaningful at all must be determined in future studies on this subject matter.

Abbreviations

1, 2, 3	first, second, third person
ABL	ablative
C	consonant
CAUS	causative
CNT	continuative
COMM	common-noun marker
DEM	demonstrative
DIR	directional
DIS	distal
DU	dual
ERG	ergative
EXCL	exclusive
GEN	genitive
IMPF	imperfective
INCL	inclusive
INDEF	indefinite
INS	instrumental
LIM	limiter
LINK	linker particle
LOC	locative
MAN	manner
MID	mid distance
NP	noun phrase
PAC	paucal
PASS	passive
PC	possessive classifier
PERSN	personal name
PL	plural
PLACEN	place name
PP	prepositional phrase
PREP	preposition
PRO	pronoun
PROP	proprial marker
PROX	proximate
PT	past tense
REAL	realis
RED	reduplication

SAG	Special Anthroponymic Grammar
SG	singular
SOG	Special Onymic Grammar
STG	Special Toponymic Grammar
SUB	subordinator
TMA	tense/mood/aspect marker
TOP	topic
TR	trial
TRANS	transitive

References

Alderete, John. 1998. Canonical types and noun phrase configuration in Fijian. In Matt Pearson (ed.), *UCLA Occasional Papers in Linguistics 21: Recent papers in Austronesian linguistics*, 19–44. Los Angeles: University of California at Los Angeles.

Aranovich, Raúl. 2013. Transitivity and polysynthesis in Fijian. *Language* 89(3). 465–500.

Bauer, Winifred. 1993. *Maori*. London & New York: Routledge.

Caro Reina, Javier. 2014. The grammaticalization of the terms of address *en* and *na* as onymic markers in Catalan. In Friedhelm Debus, Rita Heuser & Damaris Nübling (eds.), *Linguistik der Familiennamen* (Germanistische Linguistik 225–227), 175–204. Hildesheim: Olms.

Caro Reina, Javier. 2020. The definite article with place names in Romance languages. In Nataliya Levkovych & Julia Nintemann (eds.), *Aspects of the grammar of names: Empirical case studies and theoretical topics* (LINCOM Studies in Language Typology 33), 25–52. München: LINCOM.

Caro Reina, Javier. this volume. The definite article with personal names in Romance languages.

D'hulst, Yves, Rolf Thieroff & Trudel Meisenburg. this volume. River names: Definite articles and geographical names.

Dixon, R.M.W. 1988. *A grammar of Boumaa Fijian*. Chicago & London: The University of Chicago Press.

Geraghty, Paul. 1976. Fijian prepositions. *The Journal of the Polynesian Society* 85(4). 507–520.

Geraghty, Paul. 2011. Nadrogā. In John Lynch, Malcolm Ross & Terry Crowley (eds.), *The Oceanic languages*, 833–847. London & New York: Routledge.

Handschuh, Corinna. 2017. Nominal category marking on personal names: A typological study of case and definiteness. In Tanja Ackermann & Barbara Schlücker (eds.), *The morphosyntax of proper names*. [Special issue]. *Folia Linguistica* 51(2). 483–504.

Handschuh, Corinna. 2018. *Distinct marking of common and proper nouns in Oceanic (and beyond). Synchronic variation in form and function and historical implications*. Paper presented at the *14th International Conference on Austronesian Languages* (17–20 July, 2018, Antananarivo/Madagascar).

Handschuh, Corinna. 2019. The classification of personal names: A crosslinguistic study of sex-specific forms, classifiers and gender marking on personal names. In Antje Dammel &

Corinna Handschuh (eds.), *Grammar of names*. [Special issue]. *Language Typology and Universals* 72(4). 539–572.

Handschuh, Corinna & Antje Dammel. 2019. Introduction: Grammar of names and grammar out of names. In Antje Dammel & Corinna Handschuh (eds.), *Grammar of names*. [Special issue]. *Language Typology and Universals* 72(4). 453–466.

Heidenkummer, Alexandra & Johannes Helmbrecht. 2017. Form, Funktion und Grammatikalisierung des Eigennamenmarkers =ga im Hoocąk (Sioux). In Johannes Helmbrecht, Damaris Nübling & Barbara Schlücker (eds.), *Namengrammatik* (Linguistische Berichte – Sonderheft 23), 11–32. Hamburg: Buske.

Helmbrecht, Johannes. this volume. Proper names with and without definite articles: preliminary results.

Himmelmann, Nikolaus P. 2005. The Austronesian languages of Asia and Madagascar: Typological characteristics. In Alexander Adelaar & Nikolaus P. Himmelmann (eds.), *The Austronesian languages of Asia and Madagaskar*, 110–181. London: Routledge.

Hockett, Charles F. 1958. *A course in modern linguistics*. New York: Macmillan.

Lynch, John. 1998. *Pacific languages: An introduction*. Honolulu: University of Hawai'i Press.

Lynch, John, Malcolm Ross & Terry Crowley. 2011. Typological overview. In John Lynch, Malcolm Ross & Terry Crowley (eds.), *The Oceanic languages*, 34–53. London & New York: Routledge.

Nübling, Damaris. 2005. Zwischen Syntagmatik und Paradigmatik: Grammatische Eigennamenmarker und ihre Typologie. *Zeitschrift für Germanistische Linguistik* 33(1). 25–56.

Nübling, Damaris. 2017a. The growing distance between proper names and common nouns in German: On the way to onymic schema constancy. In Tanja Ackermann & Barbara Schlücker (eds.), *The morphosyntax of proper names*. [Special issue]. *Folia Linguistica* 51(2). 341–367.

Nübling, Damaris. 2017b. Funktionen neutraler Genuszuweisung bei Personennamen und Personenbezeichnungen im germanischen Vergleich. In Johannes Helmbrecht, Damaris Nübling & Barbara Schlücker (eds.), *Namengrammatik* (Linguistische Berichte – Sonderheft 23), 173–211. Hamburg: Buske.

Nübling, Damaris, Fabian Fahlbusch & Rita Heuser. 2015. *Namen: Eine Einführung in die Onomastik*. 2nd edn. Tübingen: Narr.

Reid, Lawrence A. 2002. Determiners, nouns, or what? Problems in the analysis of some commonly occurring forms in Philippine languages. *Oceanic Linguistics* 41(2). 295–309.

Ross, Malcolm. 2011. Taiof. In John Lynch, Malcolm Ross & Terry Crowley (eds.), *The Oceanic languages*, 426–439. London & New York: Routledge.

Salaberri, Iker. this volume. D-marking on Basque personal names from a synchronic and diachronic perspective.

Schütz, Albert J. 1985. *The Fijian language*. Honolulu: University of Hawai'i Press.

Schütz, Albert J. 2014. *Fijian reference grammar*. Honolulu: University of Hawai'i Press.

Schütz, Albert J. & Rusiate T. Komaitai. 1971. *Spoken Fijian*. Honolulu: University of Hawai'i Press.

Schütz, Albert J. & Tevita Nawadara. 1972. A refutation of the notion 'passive' in Fijian. *Oceanic Linguistics* 11(2), 88–109.

Seiler, Hansjakob. 1986. *Apprehension. Language, object, and order. Part III: The universal dimension of apprehension*. Tübingen: Narr.

Stolz, Thomas. 2020. Is there something wrong with *iya*? On morphosyntactic issues connected to place names in Chamorro. In Nataliya Levkovych & Julia Nintemann (eds.), *Aspects of the grammar of names: Empirical case studies and theoretical topics* (LINCOM Studies in Language Typology 33), 53–146. München: LINCOM.

Stolz, Thomas, Nataliya Levkovych & Aina Urdze. 2017a. Die Grammatik der Toponyme als typologisches Forschungsfeld. In Johannes Helmbrecht, Damaris Nübling & Barbara Schlücker (eds.), *Namengrammatik* (Linguistische Berichte – Sonderheft 23), 121–146. Hamburg: Buske.

Stolz, Thomas, Nataliya Levkovych & Aina Urdze. 2017b. When zero is just enough … In support of *Special Toponymic Grammar* in Maltese. In Tanja Ackermann & Barbara Schlücker (eds.), *The morphosyntax of proper names*. [Special issue]. *Folia Linguistica* 51(2). 453–482.

Stolz, Thomas, Nataliya Levkovych & Aina Urdze. 2018. La morfosintassi dei toponimi in prospettiva tipologica. In Giuseppe Brincat & Sandro Caruana (eds.), *Tipologia e 'dintorni': il metodo tipologico alla intersezione di piani d'analisi*, 307–324. Roma: Bulzoni.

Topping, Donald M. & Bernadita C. Dungca. 1973. *Chamorro reference grammar*. Honolulu: University of Hawaiʻi Press.

van Urk, Coppe. 2020. Object licensing in Fijian and the role of adjacency. *Natural Language and Linguistic Theory* 38. 313–364.

Whorf, Benjamin Lee. 1945. Grammatical categories. *Language* 21(1). 1–11.

www.ingramcontent.com/pod-product-compliance
Lightning Source LLC
Chambersburg PA
CBHW050520170426
43201CB00013B/2021